Hunting Causes and Using Them

Hunting Causes And Using Them argues that causation is not one thing, as commonly assumed, but many. There is a huge variety of causal relations, each with different characterizing features, different methods for discovery and different uses to which it can be put. In this collection of new and previously published essays, Nancy Cartwright provides a critical survey of philosophical and economic literature on causality, with a special focus on the currently fashionable Bayes-nets and invariance methods – and exposes a huge gap in that literature. Almost every account treats either exclusively of how to hunt causes or of how to use them. But where is the bridge between? It's no good knowing how to warrant a causal claim if we don't know what we can do with that claim once we have it.

This book is for philosophers, economists and social scientists – or for anyone who wants to understand what causality is and what it is good for.

NANCY CARTWRIGHT is Professor of Philosophy at the London School of Economics and Political Science and at the University of California, San Diego, a Fellow of the British Academy and a recipient of the MacArthur Foundation Award. She is author of *How the Laws of Physics Lie* (1983), *Nature's Capacities and their Measurement* (1989), *Otto Neurath: Philosophy Between Science and Politics* (1995) with Jordi Cat, Lola Fleck and Thomas E. Uebel, and *The Dappled World: A Study of the Boundaries of Science* (1999).

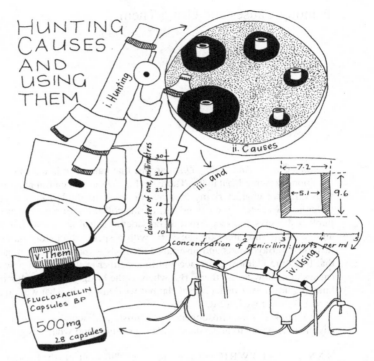

HUNTING CAUSES AND USING THEM

Drawing by Rachel Hacking Gee

University of Oxford's Museum of the History of Science:

Lord Florey's team investigated antibiotics in 1939. They succeeded in concentrating and purifying penicillin. The strength of penicillin preparations was determined by measuring the extent to which it prevented bacterial growth. The penicillin was placed in small cylinders and a culture dish and the size of the clear circular inhibited zone gave an indication of strength. Simple apparatus turned this measurement into a routine procedure. The Oxford group defined a standard unit of potency and was able to produce and distribute samples elsewhere.

A specially designed ceramic vessel was introduced to regularize penicillin production. The vessels could be stacked for larger-scale production and readily transported. The vessels were tipped up and the culture containing the penicillin collected with a pistol. The extraction of the penicillin from the culture was partly automated with a counter-current apparatus. Some of the work had to be done by hand using glass bottles and separation funnels.

Penicillin was obtained in a pure and crystalline form and used internationally.

Hunting Causes and Using Them

Approaches in Philosophy and Economics

Nancy Cartwright

CAMBRIDGE
UNIVERSITY PRESS

University Printing House, Cambridge CB2 8BS, United Kingdom

Cambridge University Press is part of the University of Cambridge.

It furthers the University's mission by disseminating knowledge in the pursuit of education, learning and research at the highest international levels of excellence.

www.cambridge.org
Information on this title: www.cambridge.org/9780521860819

© Nancy Cartwright 2007

First published 2007

A catalogue record for this publication is available from the British Library

ISBN 978-0-521-86081-9 Hardback
ISBN 978-0-521-67798-1 Paperback

For Lucy

Contents

Acknowledgements

Very many people have helped over the years with the work in this volume and I am extremely grateful to them. Specific references will be found in each of the separate chapters. More generally the recent work and the overarching outline for the volume owe much to Julian Reiss and Damien Fennell. Gabriele Contessa, Damien Fennell, Dorota Rejman and Sheldon Steed, all from the London School of Economics (LSE), plus many people at Cambridge University Press worked hard in the production of the volume and Sheldon Steed on the index. Discussions in seminars at LSE and the University of California at San Diego have pushed the work forward considerably and I would like to thank students and colleagues in both places for their help and especially the work of Julian Reiss, who has contributed much to my thinking on causality. Pat Suppes, Ruth Marcus and Adolf Grunbaum have always stood over my shoulder, unachievable models to be emulated, as has Stuart Hampshire of course, who always thought my interest in social science was a mistake. Special thanks are due to Rachel Hacking Gee for the cover drawing.

Funding for the work has been provided from a number of sources which I wish to thank for their generosity and support. The (UK) Arts and Humanities Research Board supported the project Causality: Metaphysics and Methods. The British Academy supported trips to Princeton's Center for Health and Wellbeing to work with economist Angus Deaton on causal inference about the relations between health and status, which I also studied with epidemiologist Michael Marmot. I have had a three-year grant from the Latsis Foundation to help with my research on causality and leave-time supported by the (US) National Science Foundation under grant No. 0322579. (Any opinions, findings and conclusions or recommendations expressed in this material are those of the author and do not necessarily reflect the view of the National Science Foundation.) The volume was conceived and initiated while I was at the Center for Health and Wellbeing and the final chapters were written while I was at the Institute for Advanced Study in Bologna, where I worked especially with Maria Carla Galavotti.

Thanks to all for the help!

The author acknowledges permission to use previously published and forth-coming papers in this volume. About one-third of the chapters are new. The provenance of the others is as follows:

Chapter 2: *Philosophy of Science*, 71, 2004 (pp. 805–19).

Chapter 4: *Journal of Philosophy*, III (2), 2006 (pp. 55–66).

Chapter 6: *The Monist*, 84, 2001 (pp. 242–64). This version is from *Probability Is the Very Guide of Life*, H. Kyburg and M. Thalos (eds.), Open Court, Chicago and La Salle, Illinois, 2003 (pp. 253–76).

Chapter 7: *Stochastic Causality*, D. Costantini, M. C. Galavotti and P. Suppes (eds.), Stanford, CA, CSLI Publications, 2001 (pp. 65–84).

Chapter 8: *British Journal for the Philosophy of Science*, 53, 2002 (pp. 411–53).

Chapter 9: *British Journal for the Philosophy of Science*, 57, 2006 (pp. 197–218).

Chapter 10: *Philosophy of Science*, 70, 2003 (pp. 203–24).

Chapter 12: *Journal of Econometrics*, 67, 1995 (pp. 47–59).

Chapter 15: Discussion Paper Series, Centre for the Philosophy of Natural and Social Science, London LSE, 1999 (pp. 1–11). This version is in *The 'Experiment' in the History of Economics*, P. Fontaine and R. Leonard (eds.), London, Routledge, 2005, ch. 6.

Chapter 16: To appear in *Explanation and Causation: Topics in Contemporary Philosophy*, M. O'Rourke *et al.* (eds.), vol. IV, Boston, Mass., MIT Press, forthcoming.

Introduction

Look at what economists are saying. 'Changes in the real GDP unidirectionally and significantly Granger cause changes in inequality.'[1] Alternatively, 'the evolution of growth and inequality must surely be the outcome of similar processes' and 'the policy maker . . . needs to balance the impact of policies on both growth and distribution'.[2] Until a few years ago claims like this – real causal claims – were in disrepute in philosophy and economics alike and sometimes in the other social sciences as well. Nowadays causality is back, and with a vengeance. That growth causes inequality is just one from a sea of causal claims coming from economics and the other social sciences; and methodologists and philosophers are suddenly in intense dispute about what these kinds of claims can mean and how to test them. This collection is for philosophers, economists and social scientists or for anyone who wants to understand what causality is, how to find out about it and what it is good for.

If causal claims are to play a central role in social science and in policy – as they should – we need to answer three related questions about them:

What do they mean?

How do we confirm them?

What use can we make of them?

The starting point for the chapters in this collection[3] is that these three questions must go together. For a long time we have tended to leave the first to the philosopher, the second to the methodologist and the last to the policy consultant. That, I urge, is a mistake. Metaphysics, methods and use must march hand in hand. Methods for discovering causes must be legitimated by showing that they are good ways for finding just the kinds of things that causes are; so too the conclusions we want to draw from our causal claims, say for planning and policy, must be conclusions that are warranted given our account of what causes are. Conversely, any account of what causes are that does not dovetail with what we take to be our best methods for finding them or the standard

---

[1] Assane and Grammy (2003), p. 873.   [2] Lundberg and Squire (2003), p. 326.
[3] Many of these chapters have previously been published; about one-third are new. For original places of publication, see the acknowledgements.

uses to which we put our causal claims should be viewed with suspicion. Most importantly –

Our philosophical treatment of causation must make clear why the methods we use for testing causal claims provide good warrant for the uses to which we put those claims.

I begin this book with a defence of causal pluralism, a project that I began in *Nature's Capacities and their Measurement*,[4] which distinguishes three distinct levels of causal notions, and continued in the discussions of causal diversity in *The Dappled World*.[5] Philosophers and economists alike debate what causation is and, correlatively, how to find out about it. Consider the recent *Journal of Econometrics* volume on the causal story behind the widely observed correlations between bad health and low status. The authors of the lead article,[6] Adams, Hurd, McFadden, Merrill and Ribeiro, test the hypothesis that socio-economic status causes health by a combination of the two methods I discuss in part II: Granger causality, which is the economists' version of the probabilistic theory of causality that gives rise to Bayes-nets methods, and an invariance test. Of the ten papers in the volume commenting on the Adams et al. work, only one discusses the implementation of the tests. The other nine quarrel with the tests themselves, each offering its own approach to how to characterize causality and how to test for it.

I argue that this debate is misdirected. For the most part the approaches on offer in both philosophy and economics are not alternative, incompatible views about causation; they are rather views that fit different kinds of causal systems. So the question about the choice of method for the Adams et al. paper is not 'What is the "right" characterization of causality?' but rather, 'What kind of a causal system is generating the AHEAD (Asset and Health Dynamics of the Oldest Old) panel data that they study?'

Causation, I argue, is a highly varied thing. What causes should be expected to do and how they do it – really, what causes are – can vary from one kind of system of causal relations to another and from case to case. Correlatively, so too will the methods for finding them. Some systems of causal relations can be regimented to fit, more or less well, some standard pattern or other (for example, the two I discuss in part II) – perhaps we build them to that pattern or we are lucky that nature has done so for us. Then we can use the corresponding method from our tool kit for causal testing. Maybe some systems are idiosyncratic. They do not fit any of our standard patterns and we need system-specific methods to learn about them. The important thing is that there is no single interesting characterizing feature of causation; hence no off-the-shelf or one-size-fits-all method for finding out about it, no 'gold standard' for judging causal relations.[7]

---

[4] Cartwright (1989).   [5] Cartwright (1999).   [6] Adams et al. (2003).
[7] See John Worrall (2002) on why randomized clinical trials are not a gold standard.

Part II illustrates this with two different (though related) kinds of causal system, matching two different philosophical accounts of what causation is, two different methodologies for testing causal claims and two different sets of conclusions that can be drawn once causal claims are accepted.

The first are systems of causal relations that can be represented by causal graphs plus an accompanying probability measure over the variables in the graph. The underlying metaphysics is the probabilistic theory of causality, as first developed by Patrick Suppes. The methods are Bayes-nets methods. Uses are licensed by a well-known theorem about what happens under 'intervention' (which clearly needs to be carefully defined) plus the huge study of the counterfactual effects of interventions by Judea Pearl. I take up the question of how useful these counterfactuals really are in part III.

In part II, I ask 'What is wrong with Bayes nets?' My answer is really, 'nothing'. We can prove that Bayes-nets methods are good for finding out about systems of causal relations that satisfy the associated metaphysical assumptions. The mistake is to suppose that they will be good for all kinds of systems. Ironically, I argue, although these methods have their metaphysical roots in the probabilistic theory of causality, they cannot be relied on when causes act probabilistically. Bayes-nets causes must act deterministically; all the probabilities come from our ignorance. There are other important restrictions on the scope of these methods as well, arising from the metaphysical basis for them. I focus on this one because it is the least widely acknowledged.

The second kind of system illustrated in part II is systems of causal relations that can be represented by sets of simultaneous linear equations satisfying specific constraints. The concomitant tests are invariance tests. If an equation represents the causal relations correctly, it should continue to obtain (be invariant) under certain kinds of intervention. This is a doctrine championed in various forms by both philosophers and economists. On the philosophical side the principal advocates are probably James Woodward and Daniel Hausman; for economics, see the paper on health and status mentioned above or econometrician David Hendry, who argues that causes must be *super*exogenous – they must satisfy certain probabilistic conditions (exogeneity conditions) and they must continue to do so under the policy interventions envisaged. (I discuss Hendry's views further in chs. 4 and 16.)

My discussion in part II both commends and criticizes these invariance methods. In praise I lay out a series of axioms that makes their metaphysical basis explicit. The most important is the assumption of the priority of causal relations, that causal relations are the 'ontological basis' for all functionally true relations, plus some standard assumptions (like irreflexivity) about causal order. 'Two theorems on invariance and causality' first identifies a reasonable sense of 'intervention' and a reasonable definition of what it means for an equation to 'represent the causal relations correctly' and then proves that the methods are matched to the metaphysics. Some of the uses supported by this kind of causal metaphysics are described in part I.

As with Bayes nets, my criticisms of invariance methods come when they overstep their bounds. One kind of invariance at stake in this discussion sometimes goes under the heading 'modularity': causal relations are 'modular' – each one can be changed without affecting the others. Part II argues that modularity can – and generally does – fail.

I focus on these two cases because they provide a model of the kind of work I urge that we should be doing in studying causation. Why is it that I can criticize invariance or Bayes-nets methods for overstepping their bounds? Because we know what those bounds are. The metaphysical theories tell us what kinds of system of causal relations the methods suit, and both sides – the methods and the metaphysics – are laid out explicitly enough for us to show that this is the case. The same too with the theorems on use. This means that we know (at least 'in principle') when we can use which methods and when we can draw which conclusions.

Part III of this book looks at a number of economic treatments of causality. The chapter on models and Galilean experiments simultaneously tackles causal inference and another well-known issue in economic methodology, 'the unrealism of assumptions' in economic models. Economic models notoriously make assumptions that are highly unrealistic, often 'heroic', compared to the economic situations that they are supposed to treat. I argue that this need not be a problem; indeed it is necessary for one of the principal ways that we use models to learn about causes.

Many models are thought experiments designed to find out what John Stuart Mill called the 'tendency' of a causal factor – what it contributes to an outcome, not what outcomes will actually occur in the complex world where many causes act together. For this we need exceptional circumstances, ones where there is nothing else to interfere with the operation of the cause in producing its effect, just as with the kinds of real experiment that Galileo performed to find out the effects of gravity. My discussion though takes away with one hand what it gives with the other. For not all the unrealistic assumptions will be of this kind. In the end, then, the results of the models may be heavily overconstrained, leading us to expect a far narrower range of outcomes than those the cause actually tends to produce.

The economic studies discussed in part III themselves illustrate the kind of disjointedness that I argue we need to overcome in our treatment of causality. Some provide their own accounts of what causation is (economist/methodologist Kevin Hoover and economists Steven LeRoy and David Hendry); others, how we find out about it (Herbert Simon as I reconstruct him and my own account of models as Galilean experiments); others still, what we can do with it (James Heckman and Steven LeRoy on counterfactuals). The dissociation can even come in the interpretation of the same text. Kevin Hoover (see ch. 14, 'The merger of cause and strategy: Hoover on Simon on causation') presents his account as a generalization to non-linear systems of Herbert Simon's characterization of causal order in linear systems. My 'How to get causes from probabilities: Cartwright on Simon on causation' (ch. 13) provides a different story of what Simon might have been doing. The chief difference is that I focus on how we confirm causal claims, Hoover on what use they are to us.

The turn to economics is very welcome from my point of view because of the focus on use. In the triad metaphysics, methods and use, use is the poor sister in philosophic accounts of causality. Not so in economics, where policy is the point. This is why David Hendry will not allow us to call a relation 'causal' if it slips away in our fingers when we try to harness it for policy. And Hoover's underlying metaphysics is entirely based on the demand that we must be able to use causes to bring about effects.

Perhaps it seems an unfair criticism of our philosophic accounts to say they are thin on use. After all one of our central philosophic theories equates causality with counterfactuals and another equates causes with whatever we can manipulate to produce or change the effect. Surely both of these provide immediate conclusions that help us figure out which policies and techniques will work and which not? I think not. The problem is one we can see by comparing Hoover's approach to Simon with mine. What we need is to join the two approaches in one, so that we simultaneously know how to establish a causal claim and what use we can make of that claim once it is established.

Take counterfactuals first. The initial David Lewis style theory[8] takes causal claims to be tantamount to counterfactuals: C causes E just in case if C had not occurred, E would not have occurred. Recent work looks at a variety of different causal concepts – like 'prevents', 'inhibits' or 'triggers' – and provides a different counterfactual analysis of each.[9] The problem is that we have one kind of causal claim, one kind of counterfactual. If we know the causal claim, we can assert the corresponding counterfactual; if we know the counterfactual, we can assert the corresponding causal claim. But we never get outside the circle.

The same is true of manipulation accounts. We can read these accounts as theories of what licenses us to assert a causal claim or as theories that license us to infer that when we manipulate a cause, the effect will change. We need a theory that does both at once. Importantly it must do so in a way that is both justified and that we can apply in practice.

---

[8] Lewis (1973).   [9] Hall and Paul (2003).

This brings me to the point of writing this book. In studying causality, there are two big jobs that face us now:

> Warrant for use: we need accounts of causality that show how to travel from our evidence to our conclusions. Why is the evidence that we take to be good evidence for our causal claims good evidence for the conclusions we want to draw from these claims? In the case of the two kinds of causal system discussed in part II, it is metaphysics – the theory of probabilistic causality for the first and the assumption of causal priority for the second – that provides a track from method to use. That is the kind of metaphysics we need.

> Let's get concrete: our metaphysics is always too abstract. That is not surprising. I talk here in the introduction loosely about the probabilistic theory of causality and causal priority. But loose talk does not support proofs. For that we need precise notions, like 'the causal Markov condition', 'faithfulness' and 'minimality'. These tell us exactly what a system must be like to license Bayes-nets methods for causal inference and Bayes-nets conclusions. What do these conditions amount to in the real world? Are there even rough identifying features that can give us a clue that a system we want to investigate satisfies these abstract conditions? In the end even the best metaphysics can do no work for us if we do not know how to identify it in the concrete.

By the end of the book I hope the reader will have a good sense of what these jobs amount to and of why they are important. I hope some will want to try to tackle them.

Part I

Plurality in causality

1   Preamble

The title of this part is taken from Maria Carla Galavotti.[1] Galavotti, like me, argues that causation is a highly varied thing. There are, I maintain, a variety of different kinds of relations that we might be pointing to with the label 'cause' and each different kind of relation needs to be matched with the right methods for finding out about it as well as with the right inference rules for how to use our knowledge of it.[2]

Chapter 2, 'Causation: one word, many things', defends my pluralist view of causality and suggests that the different accounts of causality that philosophers and economists offer point to different features that a system of particular causal relations might have, where the relations themselves are more precisely described with thick causal terms – like 'pushes', 'wrinkles', 'smothers', 'cheers up' or 'attracts' – than with the loose, multi-faceted concept *causes*. It concludes with the proposal that labelling a specific set of relations 'causal' in science can serve to classify them under one or another well-known 'causal' scheme, like the Bayes-nets scheme or the 'structural' equations of econometrics, thus warranting all the conclusions about that set of relations appropriate to that scheme.

Whereas ch. 2 endorses an ontological pluralism, ch. 3, 'Causal claims: warranting them and using them', is epistemological. It describes the plurality of methods that can provide warrant for a causal conclusion. It is taken from a talk given at a US National Research Council conference on evidence in the social sciences and for social policy, in response to the drive for the hegemony of the randomized controlled trial. There is a huge emphasis nowadays on evidenced-based policy. That is all to the good. But this is accompanied by a tendency towards a very narrow view of what counts as evidence.

In many areas it is taken for granted that by far the best – and perhaps the only good – kind of evidence for a policy is to run a pilot study, a kind of mini version of the policy, and conduct a randomized controlled trial to evaluate the

---

[1] Galavotti (2005).
[2] See Cat (forthcoming) for a discussion of different kinds of causality which apply to diverse cases in the natural, social and medical sciences.

effectiveness of the policy in the pilot situation. All other kinds of evidence tend to be ignored, including what might be a great deal of evidence that suggested the policy in the first place.

This is reminiscent of a flaw in reasoning that Daniel Kahneman and Amos Tversky[3] famously accuse us all of commonly making, the neglect of base rate probabilities in calculating the posterior probability of an event. We focus, they claim, on the conditional probability of the event and neglect to weigh in the prior probability of the event based on all our other evidence. It is particularly unfortunate in studies of social policy because of the well-known difficulties that face the randomized controlled trial at all stages, like the problem of operational-izing and measuring the desired outcome, the comparability of the treatment and control groups, pre-selection, the effect of having some policy at all, the effects of the way the policy is implemented, the similarity of the pilot situation to the larger target situation and so on.

Chapter 3 is so intent on stressing the plurality of methods for claims of causality and effectiveness that it neglects the ontological pluralism argued for in ch. 2. This neglect is remedied in ch. 4. If we study a variety of different kinds of causal relations in our sciences then we face the task of ensuring that the methods we use on a given occasion are appropriate to the kind of relation we are trying to establish and that the inferences we intend to draw once the causal claims are established are warranted for that kind of relation. This is just what we could hope our theories of causality would do for us. 'Where is the theory in our "theories" of causality?' suggests that they fail at this. This leaves us with a huge question about the joint project of hunting and using causes: what is it about our methods for causal inference that warrants the uses to which we intend to put our causal results?

---

[3] Kahneman and Tversky (1979).

2　Causation: one word, many things

2.1　Introduction

I am going to describe here a three-year project on causality under way at the London School of Economics (LSE) funded by the British Arts and Humanities Research Board. The central idea behind my contribution to the project is Elizabeth Anscombe's.[1] My work thus shares a lot in common with that of Peter Machamer, Lindley Darden and Carl Craver, which is also discussed at these Philosophy of Science Association meetings. My basic point of view is adumbrated in my 1999 book *The Dappled World*:[2]

> The book takes its title from a poem by Gerard Manley Hopkins. Hopkins was a follower of Duns Scotus. So too am I. I stress the particular over the universal and what is plotted and pieced over what lies in one gigantic plane . . .

> About causation I argue . . . there is a great variety of different kinds of causes and that even causes of the same kind can operate in different ways . . .

> The term 'cause' is highly unspecific. It commits us to nothing about the kind of causality involved nor about how the causes operate. Recognizing this should make us more cautious about investing in the quest for universal methods for causal inference.

The defence of these claims proceeds in three stages.

Stage 1: as a start I shall outline troubles we face in taking any of the dominant accounts now on offer as providing universal accounts of causal laws:[3]

1　the probabilistic theory of causality (Patrick Suppes) and consequent Bayes-nets methods of causal inference (Wolfgang Spohn, Judea Pearl, Clark Glymour);
2　modularity accounts (Pearl, James Woodward, economist Stephen LeRoy);
3　the invariance account (Woodward, economist/philosopher Kevin Hoover);
4　natural experiments (Herbert Simon, Nancy Cartwright);

---

[1] Anscombe (1993 [1971]).　　[2] Cartwright (1999), ch. 5.

[3] I exclude the counterfactual analysis of causation from consideration here because it is most plausibly offered as an account of singular causation. At any rate, the difficulties that the account faces are well known.

5 causal process theories (Wesley Salmon, Phil Dowe);
6 the efficacy account (Hoover).

Stage 2: if there is no universal account of causality to be given, what licenses the word 'cause' in a law? The answer I shall offer is: thick causal concepts.

Stage 3: so what good is the word 'cause'? Answer: that depends on the assumptions we make in using it – hence the importance of formalization.

2.2   Dominant accounts of causation

The first stage is the longest. It involves a review of what I think are currently the most dominant accounts of causal laws that connect with practical methods. Let us just look at a few of these cases to get a sense of the kinds of things that go wrong for them. What I want to notice is a general feature of the difficulties each faces. Each account is offered with its own paradigm of a causal system and each works fairly well for its own paradigm. This is a considerable achievement – often philosophical criticism of a proposed analysis points out that the analysis does not even succeed in describing the very system offered as an exemplar. But what generally fails in the current accounts of causality on offer is that they do not succeed in treating the exemplars employed in alternative accounts.

2.2.1   Bayes-nets methods

These methods do not apply where:
1 positive and negative effects of a single factor cancel;
2 factors can follow the same time trend without being causally linked;
3 probabilistic causes produce products and by-products;
4 populations are overstratified (e.g. they are homogeneous with respect to a common effect of two factors not otherwise causally linked);
5 populations with different causal structures or (even slightly) different probability measures are mixed;
6 . . . [4]
I will add one further note to this list. Recall that the causal Markov condition, which is violated in many of the circumstances in my list, is central to Bayes nets. Advocates of Bayes-nets methods for causal inference often claim in their favour that '[a]n instance of the Causal Markov assumption is the foundation of the theory of randomized experiments'.[5]

But this cannot be true. The arguments that justify randomized experiments do not suppose the causal Markov condition; and the method works without the assumption that the populations under study satisfy the condition. Using only

---

[4] For further discussion see ch. 6.   [5] Spirles et al. (1996), p. 3.

some weaker assumptions that Bayes-nets methods also presuppose, we can prove that an ideal randomized experiment will give correct results for typical situations where the causal Markov condition fails, e.g. cases of overstratification, the probabilistic production of products and by-products, or mixing.

2.2.2   Modularity accounts

These require that each law describe a 'mechanism' for the effect, a mechanism that can vary independently of the law for any other effect. I am going to dwell on this case because it provides a nice illustration of my general thesis.

So far I have only seen discussions of modularity with respect to systems like this:[6]

$$x_1 \, c^= u_1$$
$$x_2 \, c^= f_2(x_1) + u_2$$
$$x_3 \, c^= f_3(x_1, x_2) + u_3$$
$$\cdots$$
$$x_m \, c^= f_m(x_1, \ldots, x_{m-1}) + u_m$$

where these are supposed to be causal laws for a set of quantities represented by $V = \{x_1, \ldots, x_m\}$ and where[7]

$$\forall i \, (u_i \notin V) \tag{1}$$
$$\forall i \, \neg \exists j (x_i c \to u_j, x_i \in V) \tag{2}$$

(Since u's are not caused by any quantities in V, following conventional usage I shall call the u's 'exogenous'.)

Modularity requires that it is possible either to vary one law and only one law or that each exogenous variable can vary independently of each other. So modularity implies either

(i) $x_1 \, c^= u_1$
$$x_2 \, c^= f_2(x_1) + u_2$$
$$x_3 \, c^= f_3(x_1, x_2) + u_3$$
$$\cdots$$
$$x_n \, c^= f_n(x_1, \ldots, x_{n-1}) + u_n \to x_n c^= X$$
$$x_{n+1} \, c^= f_{n+1}(x_1, \ldots, x_n) + u_{n+1}$$
$$\cdots$$

or

(ii) that there are no cross-restraints among the values of the u's.

---

[6] The symbol 'c$^=$' means that the left-hand and right-hand sides are equal and that the factors on the right-hand side are a full set of causes of the factor represented on the left.

[7] 'c c $\to$ e' means 'c is a cause of e'.

Why should systems of causal laws behave like this? Woodward's main thesis is that this kind of modularity is the (single best) marker of what it is for a set of relationships to be causal.[8] He supports this with a lot of examples, but the issues he raises are frequently ones of identifiability, which are relevant only to the epistemology of causal laws not to their metaphysics.

Hausman (1998) also takes modularity as central to the idea of causation. He adds an empirical consideration to support the fact that systems of causal laws will always be modular. Although we may tend to focus on one or two or a handful of salient causal factors, in reality the cause of any factor is always very complex. This makes it likely that any two factors will always have some components of their total cause that are unrelated to each other and that thus can be used to manipulate the two factors independently. This may be plausible in cases of singular causation (with respect to purely counterfactual manipulations) that occur outside any regimented system, but it does not seem true in systems where the causal behaviour is repeatable and the causal laws depend on a single underlying structure.

I shall illustrate this below. But first I would like to look in some detail at an argument in support of modularity that has received less attention in the philosophical literature. Judea Pearl and Stephen LeRoy[9] both make claims about ambiguity that I also find an echo of in Woodward. Causal analysis, Pearl tells us, 'deals with changes' – and here he means changes under an 'intervention' that changes *only* the cause (and anything that must change in train).[10] So

Pearl/LeRoy requirement: a causal law for the effect of x_c on x_e is supposed to state unambiguously what difference a unit change of x_c (by 'intervention') will make on x_e.

I always find it puzzling why we should think that a law for the effect of e on c should tell us what happens to e when the set of laws is itself allowed to alter or even when c is brought about in various ways. I would have thought that if there was an answer to the question, it would be contained in some other general facts – like the facts about the underlying structure that gives rise to the laws and that permits certain kinds of changes in earlier variables. My reconstruction of Pearl and LeRoy's answer to my puzzle takes them to be making a very specific claim about what a causal law is (in the kind of deterministic frameworks we have been considering):

A causal law about the effect of x_n on any other variable is Nature's instruction for determining what happens when either:

---

[8] See for instance Woodward (1997; 2000).   [9] Cooley and LeRoy (1985).
[10] Pearl (2000), p. 345 and Pearl (2002).

1 the causal law describing the causes of x_n varies from $x_n = f_n(x_1, \ldots, x_{n-1}) + u_n$ to $x_n = X$,
   or
2 the exogenous variable for x_n (i.e. u_n) varies and nothing else varies except what these variations compel.

So for every system of causal laws:
 (i) such variation in any cause must be possible, and
 (ii) the law in question must yield an unambiguous answer for what happens
     to the effect under such variation in a cause.
Hence the requirement called 'modularity'. But there must be something wrong with this conception of causal laws. When Pearl talked about this recently at LSE he illustrated this requirement with a Boolean input–output diagram for a circuit. In it, not only could the entire input for each variable be changed independently of that for each other, so too could each Boolean component of that input. But most arrangements we study are not like that. They are rather like a toaster or a laser or a carburettor.

   I shall illustrate with a casual account of the carburettor, or rather, a small part of the operation of the carburettor – the control of the amount of gas that enters the chamber before ignition. I take my account of the carburettor from David Macaulay's book, *How Things Work*.[11] Macaulay's account is entirely verbal (and this will be important to my philosophical point later on). From the verbal account we can construct the diagrammatic form that the functional laws governing the amount of gas in the chamber must take:

$$\text{gas in chamber } c^= f \text{ (airflow; } \alpha) \text{ pumped gas } + (\alpha') \tag{1}$$
$$\qquad \text{gas exiting emulsion tube}$$
$$\text{airflow } c^= g \text{ (air pressure in chamber; } \beta) \tag{2}$$
$$\text{gas exiting emulsion tube } c^= h \text{ (gas in emulsion tube, air} \tag{3}$$
$$\qquad \text{pressure in chamber; } \gamma)$$
$$\text{air pressure in chamber } c^= j \text{ (suck of the pistons, setting} \tag{4}$$
$$\qquad \text{of throttle valve; } \sigma)$$

where

$$\alpha = \alpha \text{ (geometry of chamber, } \ldots)$$
$$\alpha' = \alpha' \text{ (geometry of chamber, } \ldots)$$
$$\beta = \beta \text{ (geometry of chamber, } \ldots)$$
$$\gamma = \gamma \text{ (geometry of chamber, } \ldots)$$
$$\sigma = \sigma \text{ (geometry of chamber, } \ldots)$$

Look at equation (1). The gas in the chamber is the result of the pumped gas and the gas exiting the emulsion tube. How much each contributes is fixed by other

[11] Macaulay (1988).

factors: for the pumped gas both the amount of airflow and a parameter α, which is partly determined by the geometry of the chamber; and for the gas exiting the emulsion tube, by a parameter α', which also depends on the geometry of the chamber. The point is this. In Pearl's circuit-board, there is one distinct physical mechanism to underwrite each distinct causal connection. But that is incredibly wasteful of space and materials, which matters for the carburettor. One of the central tricks for an engineer in designing a carburettor is to ensure that one and the same physical design – for example, the design of the chamber – can underwrite or ensure a number of different causal connections that we need all at once.

Just look back at my diagrammatic equations, where we can see a large number of laws all of which depend on the same physical features – the geometry of the carburettor. So no one of these laws can be changed on its own. To change any one requires a redesign of the carburettor, which will change the others in train. By design the different causal laws are harnessed together and cannot be changed singly. So modularity fails.[12]

My conclusion though is not that we must discard modularity. Rather it is not a universal characteristic of some univocal concept of (generic) causation. There are different causal questions we can ask. We can, for instance, ask the causal question we see in the Pearl/LeRoy requirement: how much will the effect change for a unit change in the cause if the unit change in the cause were to be introduced 'by intervention'? The question will make sense and have an unambiguous answer for modular systems. The fact that many systems are not modular does not mean that this is a foolish question to ask when systems are modular.

2.2.3   Woodward's invariance account

This is a strengthening of the modularity account. Modularity accounts tell us that causal laws predict what happens under variations of the appropriate sort. Woodward's invariance account says that if a claim predicts what happens under variations of the appropriate sort, it is a causal law. Hence some of the problems for this claim are:

1 Invariance works only for systems that are modular, not for toasters and carburettors.
2 I can prove Woodward's invariance claims (once formulated explicitly) for special systems. Among the axioms for these systems are numerical transitivity, functional dependence, anti-symmetry and irreflexivity, uniqueness of coefficients, consistency and the assumption that no functional relations obtain that are not derivable from causal laws.[13] This last forbids, e.g. that two variables might show the same time trend.

So invariance also has its special problems.

---

[12] For more details see ch. 7.   [13] For full axioms, see ch. 10.

But there is one thing to note in favour of invariance methods – unlike Bayes-nets methods, they can give decisive answers about specific causal hypotheses even where the causal Markov condition fails. For instance, this is true for linear probabilistic structures like those below, where the u's serve to introduce genuine irreducible probabilities:

$$x_1 \text{ c} = b_1 + u_1$$
$$x_2 \text{ c} = a_{21}x_1 + u_2$$
$$\cdots$$
$$x_n \text{ c} = \sum_{i=1}^{n-1} a_{ni}x_i + u_n$$

In any case in which the u's are not mutually independent, the causal Markov condition will not hold. Nevertheless invariance methods will give correct judgements about individual causal hypotheses. That is, correctly formulated invariance methods will work even when the u's are correlated leading to violations of the causal Markov condition. On the other hand, because we need variations of just the right sort, where the 'right sort' is specified in causal terms, invariance methods require a great deal more specific antecedent causal knowledge than do Bayes-nets methods. Hence they are frequently of less use to us.

2.2.4   Natural experiments

If we want to tie method – really reliable method – and 'analysis' as closely as possible, probably the most natural thing would be to reconstruct our account of causality from the experimental methods we use to find out about causes.[14] Any such attempt is bound to illustrate my overall point. The conditions that must obtain for a situation to mimic that of an experiment are enormously special. A notion of causality geared to conditions that obtain in an experimental setting – whether it occurs naturally or is contrived by us – is not likely to fit well for a large variety of commonly occurring systems that other accounts (and ordinary intuitions as well) will count as causal.[15]

2.2.5   Causal processes

These accounts require that there be a continuous space–time process that conveys the causal influence from cause to effect. There is a large literature looking at the problems that arise for various specific versions of the account. But, as Kevin Hoover argues, none of them will work for crucial cases in economics that we want to study, say, cases of equilibrium, where causes and effects are

[14] Cf. Simon (1953) or Hamilton (1997). For a further discussion of these issues see Reiss (2003).
[15] For a further discussion see chs. 13 and 14 in this book.

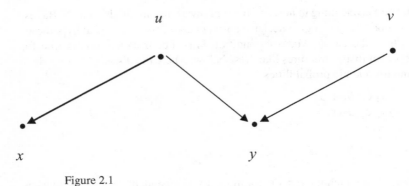

Figure 2.1

'simultaneous'; or cases involving causal relations between quantities all of which only make sense when measured over extended periods of time – which may well then overlap with each other. Hoover himself offers an account that can deal with such cases.

2.2.6   Hoover's effective strategies account

'$Xc \rightarrow Y$' if anything we do to affect X will affect Y as well, but not the reverse, maintains economist/methodologist Kevin Hoover.[16] But Hoover's characterization is too weak to serve as a universal condition on what it means for x to cause y. Consider the pattern (fig. 2.1) which we might see in a mechanical device like the toaster, where I draw the causal arrows in accord with our primitive intuitions about how the device operates – intuitions that will probably also be in accord with a causal process account of causal laws. In this case Hoover allows that x *causes* y, so long as u and v are factors that can be directly manipulated. So Hoover's condition is too weak.

On the other hand it is also too strong, since it never allows that x causes y or the reverse when the association between the two is given as pictured in fig. 2.2 (again the arrows represent causal process causality or perhaps probabilistic causation). Hence Hoover's account is too strong. Nevertheless it is based on a causal question whose answer may matter enormously to us: can we affect y by affecting x?

---

[16] See Hoover (2001) and ch. 14 here. There are also a number of well-argued 'agency' accounts in the philosophical literature. I focus on Hoover's because it is ties most closely with methodology, which is the central interest I have in finding an adequate account of causality. Also, I imagine Hoover's version of an agency account will be less familiar to philosophers of science and my discussion can provide an introduction to it.

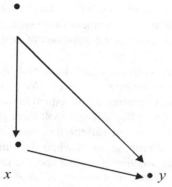

Figure 2.2

2.2.7   Diagnosis

All these accounts have problems. Does that mean that none of them is any good and we should throw them out? On the contrary, I think they all are very good. They fail, I hypothesize, because the task they set themselves cannot be accomplished. Under the influence of Hume and Kant we think of causation as a single monolithic concept. But that is a mistake. The problem is not that there are no such things as causal laws; the world is rife with them. The problem is rather that there is no single thing of much detail that they all have in common, something they share that makes them all causal laws. These investigations support a two-fold conclusion:

1 There is a variety of different kinds of causal laws that operate in a variety of different ways and a variety of different kinds of causal questions that we can ask.
2 Each of these can have its own characteristic markers; but there are no interesting features that they all share in common.

2.3   An alternative: thick causal concepts

All the accounts I described seem to suppose that there is one thing – one characteristic feature – that makes a law a causal law. I want to offer an alternative. Just as there is an untold variety of quantities that can be involved in laws, so too there is an untold variety of causal relations. Nature is rife with very specific causal laws involving these causal relations, laws that we represent most immediately using content-rich causal verbs: the pistons *compress* the air in the carburettor chamber, the sun *attracts* the planets, the loss of skill among long-term unemployed workers *discourages* firms from opening new jobs . . . These are genuine facts, but more concrete than those reported in claims that use only the abstract vocabulary of 'cause' and 'prevent'. If we overlook this, we will lose a vast amount of information that we otherwise possess, important, useful information that can help us with crucial questions of design and control.

To begin to see this alternative picture, consider again the causal equations above that describe the operation of an automobile carburettor. Where did this equation schema come from? As I said, I constructed the equations from the description of the carburettor in *How Things Work*. If you look there you will find a far more content-rich causal theory about carburettors than could be represented in equations like the ones I propose, even when the functional forms are all filled in properly. Here are some of the more specific laws that are represented by my set of causal equations. (Of course, in an engineering treatment the laws would be both quantitative and more detailed.)

1 The carburettor *feeds* gasoline and air to a car's engine . . .

2 The pistons *suck* air in though the chamber . . .

3 The low-pressure air *sucks* gasoline out of a nozzle . . .

4 The throttle valve *allows* air to flow through the nozzle . . .

5 Pressing the pedal *opens* the throttle valve more, *speeding* the airflow and *sucking in* more gasoline . . .

6 . . .

These law claims express details of the laws that govern the operation of the carburettor that are missing from the equations. If there is any doubt, just consider all the things one can learn from these kinds of thick nomological descriptions that one cannot learn from the equations. For instance, suppose we wish to increase the acceleration produced by stepping on the accelerator and we think of doing so by increasing the width of the chamber (thus allowing more gas through). Our attempt will probably be counterproductive because doing so will also affect the drop in pressure in the air as it passes through and thereby the amount of gas that can be sucked out of the nozzle.

For a Bayes-nets example, consider a case that Judea Pearl often discusses:[17]

an experiment in which soil fumigants (X) are used to increase oat crop yields (Y) by controlling the eelworm population (Z) but may also have direct effects, both beneficial and adverse, on yields beside the control of eelworms . . . farmer's choice of treatment depends on last year's eelworm population (Z_0) . . . the quantities Z_1, Z_2, and Z_3 represent, respectively, the eelworm population, both size and type, before treatment, and at the end of the season . . . B, the population of birds and other predators. (Pearl 1995, 669)

---

[17] Pearl (1995), p. 669.

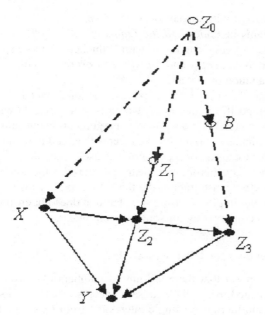

Figure 2.3 A causal diagram representing the effect of fumigants, X, on yields, Y

Variables: X: fumigants; Y: yields; B: the population of birds and other predators; Z_0: last year's eelworm population; Z_1: eelworm population before treatment; Z_2: eelworm population after treatment; Z_3: eelworm population at the end of the season.

Figure 2.3 shows the Bayes-net diagram that Pearl offers to represent the situation he describes (p. 670).

It is clear that we could give a thicker description of the causal laws operating in this experiment. Perhaps the soil fumigant *poisons* the infant eelworms, or perhaps it *smothers* the eelworm eggs, or . . .; and any of a vast number of activities could be covered by the claim that the soil fumigant has independent beneficial or adverse effects on yields. Perhaps the fumigant *enriches* the soil or *clogs* the roots. Instead Pearl gives an even thinner description. He replaces all the thick descriptions by one single piece of notation – the arrow. The arrow represents in one fell swoop all the different causal-law relations described in the thicker theory.

There is one important fact to note about thick causal concepts. They are not themselves composites from a non-causal law and some further special characteristics that make it a causal law – e.g. characteristics of the kind I have just been reviewing. Consider a comparison. Just as I contrast general causal terms like *cause* and *prevent* with thicker ones like *compress* and *attract* and *smother*, Bernard Williams in *Ethics and the Limits of Philosophy* contrasts general evaluative terms like *good* and *ought* with ' "thicker" or more specific ethical notions . . . such as *treachery* and *promise* and *brutality* and *courage*, which seem to express a union of fact and value'.[18]

But, Williams explains, they only seem to express a union of fact and value. These terms are not composites made up of two parts, a description with an evaluation added on. Elsewhere I give a whole set of arguments about causation that exactly parallels Williams's about ethical concepts.[19] Here I note only one significant point. All thick causal concepts imply 'cause'. They also imply a number of non-causal facts. But this does not mean that 'cause' + the non-causal claims + (perhaps) something else implies the thick concept. For instance we can admit that *compressing* implies *causing* + x, but that does not ensure that *causing* + x + y implies compressing for some non-circular y.

2.4   What job then does the label 'causal' do?

I have presented the proposal that there are untold numbers of causal laws, all most directly represented using thick causal concepts, each with its own peculiar truth makers; and there is no single interesting truth maker that they all share by virtue of which they are labelled 'causal' laws. What job then does the label 'causal' do?

When it comes to formal systems, we can say a lot about what job it does. That is the beauty of the formal system. The idea is that whether it is right to call something by the general term cause or not depends on what you are going to do with that label once you have attached it. Consider Pearl's work. If the causal relations, described by thick causal concepts, satisfy Pearl's modularity assumption (and if we adopt his semantics for counterfactuals), he shows a wealth of counterfactual conclusions, predictions about results of manipulations, and techniques for corroboration of specific hypotheses that we are entitled to make about these relations.

Or consider my formalizations of different versions of Woodward's invariance claims. If the Cartwright axioms are all satisfied for a given set of thick causal concepts, we can prove that an observed functional relation between quantities corresponds to a true causal claim iff the relation provides correct predictions under the right variations.

We can further prove things like the following:

1 A system of true causal-law claims including $y \stackrel{c}{=} \sum a_i x_i + u_i$ will make correct predictions about y if any of the causes of any of the x_i anywhere back in the chain is varied in the right way.

---

[18] Williams (1985), p. 129.   [19] Cartwright (2002).

2 Suppose we add assumptions that guarantee that there is a chain of causal laws between x_i and y. Then it is easy to show that if, for all i, any of the intervening factors between x_i and y vary 'to zero' in the appropriate way, y will no longer depend on x.

I also think analogous things are true even when scientific theories or claims will not bear formal reconstruction. There is still a loose set of inferences fixed by the context to which we are entitled when we make a causal-law claim with the thin word 'cause' in it. The correctness of the term 'cause' will depend on the soundness of the conclusions we draw.

To summarize, formalisms using thin causal concepts can be very useful. They provide conditions that thick causal laws might satisfy, conditions that license a specific body of inferences. General schemata using thin causal concepts are crucial for scientific practice. For they provide us with ready-made methods. Otherwise we have to find the appropriate method for each new system of laws we confront.

But there is no guarantee that we have, or can readily construct, formal schemata that will fit every system of laws we encounter. The causal arrangements of the world may be indefinitely variable. We may after all live in a dappled world.

3   Causal claims: warranting them and using them

3.1   The problem: evidence for use

Vico reminds us that it is we who have created society, so its functioning should be transparent to us. It is natural science, not social science, that should be difficult, perhaps impossible. Why then is social planning and prediction so tricky? We can build and commercially reproduce lasers so precise that complex eye surgery is routine. But we cannot build a precisely operating secondary school system. What is wrong with our knowledge in the social sciences?

Nothing is wrong with our knowledge in social science, nor with how we ascertain it, I answer. We have a panoply of methods for warranting conclusions in social science that are well tried, well developed and well understood. My hypothesis is that our problems with social policy arise primarily from the fact that we do not know how to use the knowledge we can legitimately claim to have. For good policy we need to know how to predict the consequences of very specific measures as and where they will in fact be implemented. Knowledge, whether in natural or in social science, rarely comes directly in that form; and the kinds of settings, like the auctions for the airwaves, where perhaps it does, are contra Vico, seldom ones we can (or would wish to) create. In general what we know, different pieces of knowledge of different kinds, often from a vast variety of different sources, must be brought to bear on the questions at hand. And here our methodology runs out. We are good at methods for warranting conclusions, but not for using them.

I shall defend the first part of this claim, and that is what I shall spend the bulk of my time doing, turning to use only at the end. And in keeping with

This paper was prepared for the National Research Council's conference on evidence in the social sciences and for social policy, March 2005; and for the Nordic Social Science Conference on the effects of public policy interventions, August 2005. My thanks to Damien Fennell for his help and to the National Science Foundation, the British Academy, the Latsis Foundation and the Center for Health and Wellbeing for support for the research. (The material is based upon work supported by the National Science Foundation under Grant Number 0322579. Any opinions, findings, and conclusions or recommendations expressed in this material are those of the author and do not necessarily reflect the view of the National Science Foundation.)

the concentration on knowledge that is likely to be most immediately of use in policy, I shall principally discuss methods for warranting causal claims.

3.2   Warrant

3.2.1   Two kinds of method

Methods for warranting causal claims fall into two broad categories. There are those that clinch the conclusion but are narrow in their range of application; and those that merely vouch for the conclusion but are broad in their range of application.

Derivation from theory falls into the first category, as do randomized clinical trials (RCTs), econometric methods and others. What is characteristic of methods in this category is that they are deductive: if they are correctly applied, then if the evidence claims are true, so too will the conclusions be true. That is a huge benefit. But there is an equally huge cost. These methods are concomitantly narrow in scope. The assumptions necessary for their successful application tend to be extremely restrictive and they can only take a very specialized type of evidence as input and special forms of conclusion as output.

Those in the second category – like qualitative comparative analysis (QCA) or methods that stress the importance of the mass and variety of evidence – are more wide ranging but it cannot be proved that the conclusion is assured by the evidence, either because the method cannot be laid out in a way that lends itself to such a proof or because, by the lights of the method itself, the evidence is symptomatic of the conclusion but not sufficient for it. What then is it to vouch for? That is hard to say since the relation between evidence and conclusion in these cases is not deductive and I do not think there are any good 'logics' of non-deductive confirmation, especially ones that make sense for the great variety of methods we use to provide warrant. I will say a little more about this when I catalogue a number of these methods below.

Interestingly, the method that is by far and away the most favoured by philosophers of science – the hypothetico-deductive method – straddles these two categories.

3.2.2   The straddler: the hypothetico-deductive method

Since Karl Popper and the positivists onwards, philosophers of science have taken the hypothetico-deductive method to be the one that warrants our most reliable scientific knowledge – the method by which our physics is tested. From the hypothesis under consideration in conjunction with a number of auxiliary hypotheses we deduce some more readily observable consequences. If the predicted consequences do not obtain, the hypothesis – or one of the auxiliaries – must be mistaken. This is a paradigm of a method that clinches the conclusion. If our premises are correct (premise 1:h $\rightarrow$ o; premise 2: $\neg$o) our conclusion ($\neg$h) must be correct.

But what if the predicted consequences do obtain? That is the heart of the quarrel between Popper and the positivists. Popper said that we can infer nothing; to infer that the hypothesis is true is to commit the fallacy of affirming the consequent. There is no way for a piece of evidence to distinguish between the indefinitely many hypotheses that entail it. The positivists – and the bulk of scientific practice – do not agree. They take positive results to confirm the hypothesis to some extent, then look for conditions under which the degree of confirmation would be high; for instance, if the prediction is very surprising, or very precise, or there are a great many such predictions, or the hypothesis itself is very simple, or very unifying, or . . . But none of this can turn an invalid argument into a valid one and thus provide a method that clinches the conclusion from the evidence.

I stress this because of a peculiar asymmetry. We seem to demand more of social science than of physics. We all admit that physics does pretty well. If my colleagues in philosophy of science are right, physics uses a method that cannot clinch conclusions but only vouch for them. Yet many social scientists want clinchers. I think for instance of econometricians who long for identifiability. That means that, assuming a certain abstract functional form, the probabilities inferred from the data should entail the equations of interest. We also see it frequently in discussions backing the demand for RCTs, which, as I discuss below, would be clinchers – if carried out ideally.

Of course in physics there is a rich network of knowledge and a great deal of connectedness so that any one hypothesis will have a large number of different consequences by different routes to which it is answerable. This is generally not true of hypotheses in the social sciences. My worry is that we want to use clinchers so that we can get a result from a small body of evidence rather than tackling the problems of how to handle a large amorphous body of evidence loosely connected with the hypothesis. This would be okay if only it were not for the down-side of these deductive methods – the conditions under which they can give conclusions at all are very strict.

An example of the hypothetico-deductive method at work   We find a nice example of the hypothetico-deductive method for a causal hypothesis in the work of economist Angus Deaton.[1] Deaton (like myself) does not believe in 'off-the-shelf' methodology for causal inference. Nevertheless the following example does fall under the hypothetico-deductive method.

---

[1] Conversation, November 2004, Center for Health and Wellbeing, Princeton, New Jersey.

There is a widespread correlation, revealed by different kinds of data from different populations, between low economic status and poor health. Deaton maintains that a primary source of this correlation is a causal arrow from health to income via loss of work. Unhealthy people are unable to work; this lowers their income, which is often used as a marker for status. To confirm this, Deaton looks at the National Longitudinal Mortality Study data, where there is a correlation between both low income and low education on the one hand and mortality on the other. He reasons: if the income–mortality correlation is due primarily to loss of income from poor health, then it should weaken dramatically in the retired population where health will not affect income. It should also be weaker among women than men, because the former have a weaker attachment to the labour force over the period of employment. In both cases it looks as if these predictions are borne out by the data.

Even more importantly, when the data are split between diseases that something can be done about and those that nothing can be done about, then income is correlated with mortality from both – just as it would be if causality ran from health to income. Also education is weaker or uncorrelated for the diseases that nothing can be done about. It is, he argues, hard to see how this would follow if income and education were both markers for a single concept of socio-economic status that was causal for health.

Thus the hypothesis that there is a significant causal arrow from health to income-based measures of status implies a number of specific results that seem to be borne out and that would not be expected on dominant alternative hypotheses. So the hypothesis seems to receive some confirmation – though it is very hard to say how much confirmation to award it or how far beyond the National Longitudinal Mortality Study data set to suppose it will hold. (See part III of this book on problems of exporting causal conclusions from where they are confirmed to where they will be used.)

3.2.3   Narrow methods that clinch conclusions

Derivation from theory   This is the second in rank of the philosopher's favourites. We can trust a causal conclusion that is deduced from already well-confirmed theories. This is generally supposed to be a far less useful method in the social sciences than in the natural sciences because we have no really good theories of any kind to begin with. But there are a number of factors that ameliorate this lack.

1 We need not look just to 'high' theory, abstractly expressed and systematically organized. For instance, as Naomi Oreskes argues,[2] it would be a mistake to think that we do not know the harmful effects of greenhouse gases just because

---

[2] Oreskes (forthcoming).

the results may not be derivable from this or that cutting-edge model. The basic account of radiative transfer involving CO_2 was already established in the nineteenth century, by John Tyndall, and reconfirmed by Plass and others in the twentieth century.[3] This is not 'high' theory – this is no cutting-edge climate model – but it is good science, science that has been known and accepted for a long time, based on physics theory, confirmed by laboratory experiments, etc. No one questions it, not even the climate-change deniers. So, now we go to complex climate models, 'high theory' in the sense that they are state-of-the-art, the cutting edge of the discipline. And yes, here we get a case where the details of the outcomes of increased CO_2 are uncertain because of uncertainties about the effects of other forcing functions – aerosols and clouds in particular.

On Oreskes's account what is going on in this case is a lot of fussing about the details of the predictions and, especially, about the forecasts for the future, as if one had to forecast the future to a high degree of accuracy to make a policy decision. But the fact is one often does not need a high degree of accuracy to make policy plans. One simply has to know that the basic science is well established, that it has made predictions and that those predictions are indeed coming true – a beautiful example of the hypothetico-deductive method.

2 Then there is 'common knowledge'. There is a lot that we know as well as we know anything and it is not to be disdained because it does not have the character of a 'scientific theory' or is 'merely', as Aristotle put it, knowledge of what happens 'for the most part'. 'Acorns grow into oak trees.' That is as certain as any of the surest claims of physics. Common knowledge should not be dismissed just because it is common or because we know that some of the things taken as common knowledge have turned out to be mistaken. That is characteristic of even our best scientific accounts. Just look to physics journal articles of the past. You will find a huge number of accounts of physical processes that we no longer hold with, and not just because of big theory changes like those of Newton to Einstein, but rather because that particular detailed use of the theory for that case has been superseded. The correct strategy is surely to assess the uncertainties of common knowledge, not to lose information by dismissing it or to assume that we can duck the problem of assessment by restricting ourselves to more 'scientific' claims since we face equal problems of assessing certainty there.[4]

3 We are often very clever at figuring out how to get a lot out of very little theory. Game theory methods provide one such device. The general theory

---

[3] Fleming (1998).

[4] Though perhaps not equal political problems, since these may often be less associated with differing ideologies.

supposed is exceedingly thin. Agents act so as to maximize their expected utility. Then there are auxiliary hypotheses that may or may not be met in given situations, primarily to the effect that agents can reason well and that they are informed about the structure of the 'options' and the 'pay-offs'. Then we devise specific models to consider specific causal hypotheses.

Does loss of skill among workers during periods of unemployment perpetuate periods of unemployment? One model[5] to test this hypothesis supposes that workers gain utility only from wages and leisure and entrepreneurs only from profit, that job–worker matching occurs in a specific way, that there are just two generations in the labour market, that everyone is hired/rehired at once, etc. In this model, the hypothesis can be proved true. So, assuming the theory is correct, we know that the causal claim will hold in any setting 'sufficiently' like the one described in the model. We know this with certainty since we can deduce it. The problem is to know what real situations are sufficiently like that in the model. For this we need a different kind of assessment.

Here our rigorous methodology gives out. We have rigid standards for how to 'test' results in the model but very little guidance about how to assess where the model results will apply. This is in line with my concerns about 'evidence for use' that I stress here.

Tracing the causal process (or the 'mechanisms') connecting cause and effect This method is not so common in social science as it is in more engineering-related areas, so I will not discuss it here (though it has proved important in various biological and medical studies[6]).

Probabilistic causality (Suppes or Granger causality) and the concomitant method of RCTs I want to discuss the logic of this method explicitly to underline my dual points: the logic is deductive and the argument structure is exceedingly simple; but the premises are concomitantly exceedingly strong. For both probabilistic Granger or Suppes causality and for RCTs every possible source of variation of every kind must be controlled if a valid conclusion is to be drawn.

Following the philosopher Patrick Suppes[7] and econometrician Clive Granger,[8] we suppose that for populations picked out by the right descriptions K_i, if X and Y are probabilistically dependent and X precedes Y then X causes Y. If any population P contains such a K_i as a subpopulation, then X causes Y in P in the sense that for some individuals in P, X will cause Y (in the 'long run'). This is a standard procedure in the social sciences where

[5] Pissarides (1992).   [6] Cf. Bechtel and Abrahamsen (forthcoming).
[7] Suppes (1970).   [8] Cf. Granger (1980).

we use all 'other' known causal factors to stratify a population before looking for correlations in each of the substrata as a sign of causal connections there.

The argument is deductive because of the way the K_i are supposed to be characterized. Begin from the assumption that if X and Y are probabilistically dependent in a population that must be because of the causal principles operating in that population. (Without this kind of assumption it will never be possible to establish any connection between probabilities and causality.) The trick then is to characterize the K_i in just the right way to eliminate all possible accounts of a dependency between X and Y other than that X causes Y (there is no correlation in K_i between X and any 'other' causes of Y, there is no 'selection bias', etc.). Given that K_i is specified in this way, if X and Y are probabilistically dependent in population K_i, there is no possibility left other than the hypothesis that X causes Y.

Of course the epistemic problems are enormous. How are we to know what to include in K_i? Sometimes we do know (or think we do). For instance in the Stanford Gravity Probe B experiment to test the general theory of relativity,[9] the environment is so tightly controlled that if we see the predicted precession in the gyroscopes that are now in space we can be fairly confident that nothing else could have caused them than the predicted coupling to relativistic space–time curvature.

Knowledge of just the right kind is thought to be rare in the social sciences – though we should keep in mind that it is in econometrics where we see this method in use, under the title 'Granger causality'. Granger causality solves the problem of our ignorance about just what to put in the descriptions K_i by putting in everything that happens previous to X. That of course is literally impossible so in the end very specific decisions about the nature of the K's must be made for any application.

One last thing to note about probabilistic/Granger causality is that the deductions are from probabilities to causes, not from statistics – i.e. not from summaries of data. So here is yet another source of uncertainty about the premises of the deductions. Not only might we be mistaken about the nature of the K_i for our particular system and about whether or not there can be probabilistic dependencies that have no causal source, we may also be mistaken in inferring probabilities from the data. This is a source of uncertainty that will plague any method that takes population probabilities in the premises. These include not only RCTs, Bayes-nets methods, invariance methods and methods from econometrics, but any method that looks for necessary or sufficient conditions in a population since these are just a limiting case where conditional probabilities have value 1.

[9] See Cartwright (1989), pp. 66–71.

RCTs   RCTs are designed to finesse our lack of knowledge about what other reasons might be responsible for probabilistic dependency between a treatment and an outcome. We are all familiar with this methodology so I review it exceedingly briefly. Randomization is supposed to ensure that the 'other' causal factors for Y are distributed equally in the treatment and control groups. Various blindings aim to eliminate other sources of dependency (like selection bias) and to control for factors that randomization misses.

The logic is derivative from that of probabilistic causality: if Prob(Y/X) is different in the control group from in the treatment group, it must be different in one of the K_i subpopulations;[10] and if a probabilistic dependency occurs between X and Y in a K_i subpopulation, then X must cause Y in that subpopulation and hence in any larger population of which it is a part. (This does not of course mean that it cannot also be true that X prevents Y in some other $K_j, j \neq i$, and hence prevents Y in the total population, in the same sense in which it causes Y in the total population. So, for instance, a drug that tests well in a perfectly conducted RCT will definitely be curing some group of the test population but it may simultaneously be killing those in some smaller group.)

As with any deductive method, the conclusion can only be as certain as the premises. The important point here is that by randomizing, blinding and controlling in various ways, other sources of probabilistic dependence have been eliminated or their effects calculated away. We do know some typical problems to watch out for – the placebo effect, experimenter bias and the like. But what might actually confound results in a given case requires a close and intelligent look. Confidence in the results requires that somebody knows a lot about the specific populations involved and the procedures throughout. I notice that people sometimes talk as if there is a formula for how to proceed and if we just follow it the results will be reliable. But, as with all methods, there is no avoiding the need for a great deal of good judgement, sound detailed knowledge and good sense in conducting an RCT.

Controlled experiments; natural experiments   The logic here is familiar. In principle we control so tightly that when the predicted outcome obtains, nothing but the hypothesized cause could have brought it about. It is commonplace to remark on how hard it is to do experiments in social science. But sometimes we have the good luck to find a situation in which the controls occur naturally, without our contrivance. There has been a recent push to look hard for these in order to draw causal conclusions in economics.[11] As with any deductive method, the results for either natural or contrived controlled experiments can only be as sure as our assumptions.

[10]   This is guaranteed by the fact that X and K_i will not be dependent in an ideal RCT.
[11]   Cf. Card and Krueger (1997). See also Hamilton (1997).

Bayes-nets methods   Bayes nets are graphs representing probabilistic independencies. Add some assumptions about the relations between probabilistic dependence and causality and we can use them to infer new causal relations from known causal relations and facts about probabilities – probabilities as they occur in the population under study, not experimental probabilities. The methods will produce every set of causal relations among the variables under consideration that is compatible with the input information and the background assumptions.[12]

As with the probabilistic theory of causality, Bayes-nets methods suppose that two factors will be probabilistically dependent once the 'right' background factors are held fixed if and only if they are related as cause and effect when those background factors obtain. This immediately restricts applicability; for instance the methods cannot be relied on in populations where there is 'selection bias' for joint effects. They also suppose that causes and effects will be dependent *simpliciter*, thus ruling out that the positive and negative influences of a given factor via different routes can cancel. There is in addition a kind of minimality or simplicity assumption. Importantly, as with most econometric methods for causal inference, these methods will only apply to variable sets for which the input variables (those not caused by any of the variables in the set under consideration) are all independent, which is a considerable restriction. Finally, they tell whether or not a factor is causally relevant but nothing about the strength of relevance or the functional form. (This matter is addressed in the two following methods.)

Econometric methods   Econometrics has well-developed 'structural' methods that allow the deduction of the strength of causal connections between factors in a preselected variable set, provided stringent conditions are met. These methods begin by assuming that a particular set of functional forms correctly represents the causal structure generating the observed data. What is to be discovered are the parameters that turn these functional forms into real functions – roughly, one function for each effect, where any factor that appears with a non-zero parameter on the right-hand side is judged to be a cause of that effect, with the parameter giving the strength of causal influence.[13]

In addition to assuming that the abstract functional forms are the right forms – the ones the causal principles at work actually have – the causal principles we aim to discover are also taken to meet what are called 'identification conditions'. These require that there be not too many causal connections between the factors of interest, which is necessary for disentangling the different causal

---

[12] Cf. Spirtes, Glymour and Scheines (1993).

[13] Just how these relations represent causal structure is set out by Herbert Simon and is further described in ch. 13 in this book and by Damien Fennell. See Simon (1953) and ch. 1 in Fennell (2005a).

connections from the observed data. Another important condition is that the factors taken as inputs (not caused by any factor in the preselected variable set) be probabilistically independent and also that they not restrict each other's values. This is required to guarantee that the observed data do not result from hidden common causal relationships between factors not modelled explicitly by the functional relations. Finally, statistical conditions must also be met so that the observed data sample does not 'accidentally' misrepresent the underlying data generating processes. If all of these conditions are met, then one can deduce the strength of the causal connections between factors in the variable set of interest.

Here, as with the other narrow clinching methods, secure conclusions are bought at the price of stringent conditions that are difficult to meet. In these structural methods one must know the functional form of the causal structure and that structure must not be too dense. Such conditions alone are very demanding and without them it is not clear what follows from the observed data.

Invariance methods   There is a correlation between a fall in a barometer and a storm coming. But if we manipulate the barometer in arbitrary ways (ways that vary independently from the 'other' causes of a storm), for example by smashing it, the correlation will break down. For some nice kinds of system,[14] given a sufficiently careful formulation of what we mean by invariant, we can prove that a functional relation – one we suppose we have observed to be true, say – will represent a true causal relation just in case it is invariant under all arbitrary variations of the dependent variables.

3.2.4   Broad methods that 'vouch for' conclusions

The advantage of deductive methods that clinch their conclusions is that we know exactly what we would have to do using those methods to become more certain about the conclusions – get more certain about the premises. Often we do not know how to do that; worse, frequently we know the premises are false, or probably false. These are of course problems for any kind of method, but they are especially severe for deductive methods because the requisite premises are so demanding that we cannot expect them to obtain generally. How do we know, for an RCT or Granger causality for instance, that *all* other sources of probabilistic dependence have been randomized over or controlled for and how

---

[14] For example, for linear equations where the dependent variables can take any combination of values together in systems where any true functional dependencies must result from underlying causal laws. This last is analogous to the assumption required for probabilistic causality, that all probabilistic dependencies arise from underlying causal laws. See ch. 10 in this book for details.

do we know that we are studying systems where all dependencies are due to causal connections?[15]

Here we must be careful to avoid a logical mistake. If the premises of a deductive argument are true, the conclusion must be true. What if we do not know they are true but are only willing to assign a probability to them? If we assign a probability of say 90 per cent to the premises taken jointly and we do not know anything else relevant, then it is reasonable to assign a probability of 90 per cent to the conclusion. That however is very different from the case where we are fairly certain, may even take ourselves to know, nine out of ten of the premises, but have strong reason to deny the tenth. In that case the method can make us no more certain of the conclusion than we are of that doubtful premise. Deductions can take us from truths to truths but once there is one false premise, they cannot do anything at all. That is why we need to take seriously non-deductive methods. I will review a few of these that I have worked with and try to look at what the relation of evidence to conclusion might be in each case. I will spend a little more time on the first case to exhibit the difficulty in laying out what the relation really is.

QCA[16]   This method starts from what philosophers, following J. L. Mackie,[17] call the INUS account of causation, which acknowledges both that what we usually call a cause (like C_{ij} in the formula below) is usually only a part of a total cause sufficient for the effect and that most effects have multiple separate causes. On this account causes are *i*nsufficient but *n*ecessary parts of *u*nnecessary but *s*ufficient conditions:

$$\text{INUS condition}: E \equiv C_{11}C_{12}\ldots C_{1n} \vee C_{21}C_{22}\ldots C_{2m} \vee \ldots \vee$$
$$C_{k1}C_{k2}\ldots C_{kr}{}^{[18]}$$

So to discover the causes of an effect E in a given population, sample the population, then look for factors that make a formula of INUS form true in that sample. (These methods are sometimes called 'Boolean algebra methods' because they employ huge truth tables to determine the INUS formula.) This raises the problem of statistical inference that I noted with respect to methods that move from probabilities to causes. Results in the sample may not be true of what would occur in the population as the population gets increasingly bigger.

Even in the most ideal uses, however, the method cannot clinch the results because INUS conditions are not causes. The INUS formula represents an

---

[15] Indeed we know this is frequently not the case since many factors are temporally correlated with no causal connection, so at least we had better 'detrend' data before we begin to apply the methods.

[16] See Ragin (1998). See also Lieberson (1992).   [17] Mackie (1974).

[18] Here v means 'or'.

association of features, a correlation, and we know that correlations may well be spurious. Consider for example a situation in which the following are the correct causal principles:

$$X_2 \equiv AX_1 v W$$
$$X_3 \equiv BX_1 v V$$

If these are true, so too will be

$$X_3 \equiv BX_2 \neg W v BX_1 \neg A v BX_1 A W v V$$

Thus X_2 is an INUS condition for X_3 though not a cause of it.

Suppose that we know that a given factor is an INUS condition for another, and that is all we know. Does that provide warrant for the conclusion that the first is a cause of the second; if so, how much warrant? It is not unreasonable to suppose that if a factor is a cause of another it will be an INUS condition for it; but there are many other reasons as well why it might be an INUS condition. This is just the quandary I described with the hypothetico-deductive method and it has no straightforward resolution.

Though comparative qualitative analyses cannot clinch a result, they have many advantages over various deductive methods. By contrast with RCTs and Bayes-nets methods, a QCA result is not just a yes–no verdict – 'yes, the factor is a cause', 'no, it is not'. Rather we learn the functional form of the causes. With this method we can learn that the cause is a cause for some individuals and not for others and the method is geared to determining which. Concomitantly, it is difficult to apply because it needs a complete set of causes – there is no way within the method to deal with 'omitted'/'unknown' factors as there is with econometric methods.[19]

Also, although it is not formally part of the method, the fact that we must look in detail at the individuals in the population for factors that will make up an INUS formula has great side advantages. In the first place it can alert us to a better, more concrete reformulation of the effect of interest. Very often what we aim for as an effect is something very general – improved educational attainment, better attitude, more ability to function in a job. We must operationalize these one way or another to get any study going. Looking at cases in detail often shows that our operationalization is wrong, too narrow, leaves things out, misses the mark. In the second place the choice of possible causes is more readily adjusted to the specifics of the cases at hand. The variables in the study are less likely to be standardized and hence have more flexibility to replicate the correct details of the causal stories for the individuals in the population.

---

[19] Though in econometric methods, as I noted, we often have to make very strong exogeneity assumptions about the omitted factors.

Reasoning from models and model systems   Another broad method for providing support for causal conclusions is by establishing results in a model, then reasoning from the model results to claims about the target situations. The kind of reasoning employed is as widespread and diverse as the different kinds of model used. These vary from highly concrete models, such as actual physical systems – rats or toy airplanes or prototypes – to computer simulations to extremely abstract models, such as thought experiments.

This method is used widely throughout the social and political sciences. Evolutionary models, for instance, are used to account for higher rates of death from violent causes among young men[20] or for the (currently topical) divergence in mathematical achievement between women and men.[21] Economics and political theory are rife with game theoretical models, where relatively simple premises provide persuasive hypotheses about the factors that may be driving complex phenomena. For instance[22] Schelling's model shows how segregated neighbourhoods can arise even if the individuals in those neighbourhoods individually prefer integrated neighbourhoods and Akerlof's influential 'lemons' model from microeconomics shows how asymmetric information can lead to overpayment for used cars. In social psychology ethological models are used to generate plausible hypotheses about causes of human behaviour. In medicine we use real concrete model systems, like rats. And computer simulations are gaining popularity everywhere.

In reasoning from models and model systems, two distinct questions about warrant must be answered. First, how warranted is the causal conclusion in the model? Second, how does the model conclusion provide warrant for causal claims outside the model? The first is a question of internal validity of the kind I have been considering throughout. It gets a huge variety of answers. In game theory models, the results in the model should be certain – they follow deductively. Not so in the evolutionary models where the theory is not tight enough to entail conclusions. For real model systems we have available the whole panoply of methods that we have already reviewed. The second is a question of external validity, which faces all methods since we seldom establish results in the very population and in the very situation in which we want to apply them. I turn to it when I take up issues of use.[23]

Ethnographic methods   I shall not review these since a separate review of them was made for the National Research Council conference.

Mixed indirect support   Consider Jean Perrin's influential arguments for the existence of atoms.[24] Atoms were indicated by a large number of different

---

[20] Daly and Wilson (1999).   [21] Geary (1996).   [22] These are discussed in Sugden (2000).
[23] For a more detailed discussion of internal validity in economic models see ch. 15 in this book.
[24] Discussed in Salmon (1984).

kinds of study involving different methods, in different places, with different materials, etc. Assumptions about exactly what an atom is or exactly how it behaves were not univocal across these studies.[25] Nor were any of the studies entirely satisfactory in themselves; they were almost all flawed in one way or another. Nevertheless, Perrin reasoned, atoms must exist. It would be too improbable that the different flaws in all these different kinds of studies worked out in just the right way to give the same mistaken conclusion.

Consider an analogous case in the human sciences. Recall the discussion of health and status above (see above the section '*An example of the hypothetico-deductive method at work*'). Michael Marmot[26] argues that the stress induced by low status, particularly by social isolation and high demand/low autonomy work, causes poor health. He marshals a great amount of different kinds of evidence to support the conclusion, such as long-term longitudinal studies of Whitehall civil servants, experiments on primates, statistical studies of the correlation between income and health in various places, facts and statistics about health failure in Russia, medical studies of the relations between physiological stress markers and various health problems, and psychological studies of the relationship between 'stressful' tasks and physiological stress markers.

Should this body of evidence be judged convincing? Recall that breadth, variety, precision and novelty of evidence are at the core of warrant on our most standard philosophic account of scientific method. On the hypothetico-deductive account, we look for evidence that should obtain if the hypothesis were true – then we demand that there be a lot of it, sufficiently varied, novel and not easily accounted for by other hypotheses. Figuring out if this is the case for Marmot's hypothesis – or for any hypothesis – is not easy, and it cannot be done by formula. But doing so is far more realistic than looking for some single study that could clinch a causal hypothesis like this.

3.3   Use

If warranting causal claims is a difficult matter, judging how we can put them to use is even more difficult. For it is a different enterprise altogether, requiring a different set of considerations, different kinds of background knowledge and different procedures – and these are generally far less well understood, less well articulated and less rigorous. I shall point to some of the central problems.

---

[25] Peter Galison argues that this is characteristic of contemporary physics theories. Different groups, especially theory versus experimental groups, seldom have the same interpretation for what on the face of it looks to be the same claim. See Galison (1997).

[26] Marmot (2004).

3.3.1   What claim has been established?

When we want to put our claims to use, it is essential to know exactly what claim it is that has been warranted. It is useful to think in terms of two different problems: the claim itself – what actually is established by a given method; and its scope – for what populations and in what situations the result is established.

What is the claim?   First, different methods will warrant causal claims of different forms. For instance, RCTs tell about the overall effect of a cause, averaged across subpopulations in which it may behave differently – indeed oppositely. (A drug that cures one part of the population may kill another.) Other methods require more information to apply but give more specific information. For instance, Granger causality tells what happens in each of the relevant subpopulations. What it says is that in those subpopulations (in the 'long run') the cause will produce the hypothesized effects in at least some individuals and should produce opposite effects in none. There are also well-known variations where we learn not about increased numbers of outcomes but increased levels or perhaps increases in the mean. Econometric methods give the full functional form for the relation between a set of causes and their effect; QCA also gives the full functional form, but only for yes–no variables. These are matters that we need to be alert to when we think of using the results to support policy.

A second – and age-old – problem is in deciding on the concepts to use to describe both a policy and its putative evidence. Consider an example from the natural sciences. Bodies that are not acted on by forces travel on geodesics – straight lines. But what is a straight line depends on the geometry of the surface. So suppose an experiment is performed on a sphere. The body moves in a great circle. If we take this result to be good evidence for the claim 'Bodies that are not acted on by forces move in circles', we can go far astray if the application in mind is for a flat table top.

This example illustrates that it is not always a good idea to express the conclusion that a piece of evidence is taken to warrant in too narrow or too concrete a way. On the other hand it is equally dangerous to follow the opposite strategy. The sometimes disastrous effects of overgeneralizing are well known. But also, expressing results in too abstract a vocabulary can render them almost useless, especially in social sciences where bridge laws that provide concrete interpretations of abstract concepts (like 'unemployment', 'abuse', 'incentives', etc.) are scarce. 'Love thy neighbour as thyself.' Perhaps that is good advice but what does loving one's neighbour amount to in this or that specific situation?

What is the scope of the claim?   Evidence is always collected in some population in some circumstances. With most methods the inferences that are licensed from that method are tied to the populations and situations in which the evidence is obtained and licence to go beyond those must come from somewhere outside that method.

Consider an ideal controlled experiment. It can tell us with certainty what the effect of a given cause is – in the circumstances of the experiment. But in order to do so, the circumstances of the experiment must be extremely unusual ones. What follows with 'certainty' from an ideally carried out experiment is what the cause does there, in those very unusual circumstances. The method itself tells us nothing about what the cause does elsewhere. Often the point of a controlled experiment is to establish what J. S. Mill called a tendency law.[27] These do not tell us what effect occurs when the cause is present but rather what the cause contributes to an effect in more realistic circumstances where other causally relevant factors have not been eliminated. (An example is Coulomb's law for the force exerted on one charge by another. This is never really the force a charged particle experiences in the neighbourhood of another charge since gravitational attraction will always contribute as well.)

We need three different kinds of considerations then before a causal claim from a controlled experiment can be put to use. (1) Is the experiment set up in such a way that we can conclude what we are supposed to be able to – that in the experimental situation the causal hypothesis is true? (2) Is this the kind of causal relation for which we are entitled to think there is a tendency law? On what grounds? (3) Supposing we do have a tendency law. How do we reckon what will happen in any real situation where the cause operates? For the tendency laws governing forces, we have vector addition to calculate what happens when tendencies operate together. What do we have to do the job for us in particular cases in the social sciences?

The point I want to stress is that the method of the controlled experiment, which can clinch an answer about a causal hypothesis in an experimental setting, goes no way to answering questions of the second and third type. For the most part, we have no serious methodology for answering those kinds of question, and certainly not methodologies that can be articulated and defended with the rigour with which we can treat the methods for warranting causal claims. When it comes to putting scientific claims to use, we quickly fall back on loose argument and intuition.

The issue of external validity is no less problematic for other methods. In an RCT if the population under study is 'representative' of the target population, then the results of the experiment can be extrapolated from the experimental to the target population. Here at least if we 'sample' from the target, we have good statistical guidelines for how to get a representative population. Of course we

---

[27] I discuss tendency laws in economics further in ch. 15. For cautions about drawing tendency conclusions from controlled experiments and thought experiments see Reiss (2002). See also Alexandrova (2005).

often are not able to sample from the target population, or if we can, not able to do so in the correct way. The same holds for QCA and econometric methods. Reasoning from sample models and sample systems is even more difficult. What lessons exactly are we to take away from Schelling's model about any real case of segregation? As in the case of controlled experiments, with all of these methods, rigour gives out when we try to justify exporting results from the populations and situations in which they are established. But if we cannot export results, they are of little use.

3.3.2   Are results stable under interventions?

Knowing the scope across which a result is true tells us where we can use that result for prediction.[28] But policy is more complicated. Policy involves changes: manipulating causes in hopes of producing the concomitant effects, changing them in ways they do not naturally vary as the system works on its own. Change is dangerous since we do not always know exactly what we are doing when we decide to manipulate a cause. In particular, our actions can undermine the very structure that gives rise to the causal principles we rely on to predict the outcomes of our actions.

Social scientists talk about one aspect of this problem under the heading reflexivity: people change in response to the way we study them, the way they conceive themselves, or in reaction to what they suspect will happen to them.[29]

Another aspect does not necessarily rely on the responsiveness of self-conscious agents but can arise whenever the causal principles we trust in depend on some 'deeper' 'underlying' structure. If a set of causal principles derives from some more fundamental set, then, when we change the way a cause is brought about – we bring it about in some new way by our policies – we cannot but change the underlying structure and we may well do so in a way that undermines the very principle we are trusting to for our predictions. This is a continuing theme in economics. It is the reason J. S. Mill argued that economics cannot be an inductive science;[30] econometricians have worried about it from the start;[31] and it is the basis for the famous 'Lucas critique' of macroeconomics and one of the central Chicago school arguments against government policy interventions.[32]

As before, the methods for warranting claims of stability-under-interventions for a causal connection are very different from those that warrant the causal

---

[28] We could of course think of the problems described here and in the next section as problems of the scope of a claim. But I think it is useful to divide the issues in this way since the source of the problems of scope is different in the different cases.

[29] For instance, see Finlay and Gough (2003).   [30] Mill (1836).

[31] Cf. articles by Ragnar Frisch or Trygve Haavelmo in Hendry and Morgan (1995).

[32] Lucas (1976).

claims themselves; and they are less well articulated, less well understood and less rigorous.

3.3.3 Where details matter

There may be good evidence for the effectiveness of a policy conceived, as it usually is, in the abstract, but the actual outcomes may depend crucially on the fine tuning of the method of implementation. Recall the case of laser engineering, mentioned at the beginning, and consider the early stages of development. There was a great deal of evidence, both theoretical and experimental, that 'inverted' populations of atoms properly triggered can produce highly coherent light. But we know that the results – what actually happens – depend hugely on exactly on how the laser is engineered.

Or consider poverty measures.[33] Policy may set whether a poverty line should be relative or absolute and if relative, in what way (for instance, two-thirds of the median income). But the results – for instance, the poverty ranking among European countries – depend crucially on dozens and dozens of details of implementation (how to deal with individuals versus families, wealth or welfare benefits versus earned income, etc.), details where it seems that very different decisions can be equally motivated and the rankings will come out very differently depending on how these decisions are taken.[34]

The more the details matter, the more the problems of evidence multiply. Naturally more evidence is needed to judge the consequences of taking a decision one way rather than another. But also it is unlikely that there will be much direct evidence to hand since each decision needs to be considered, not in the abstract, but in the context of the overall proposal, where the consequences of any one decision will depend on what details are supposed already to be in place. This can put severe limitations on how many alternatives can be rationally considered since working through the evidence for any one is difficult and costly. In situations like this it is important to have as good a general understanding as possible in order to make an intelligent selection of which alternatives to explore to begin with.

3.3.4 Counterfactuals and causal models

Most of our warranted causal information comes in pieces. But what we need for policy is the whole picture. We want to know what would happen if various proposed policies were implemented, and implemented in the way they would

---

[33] See Atkinson (1987; 1988).
[34] Though recall, sometimes the opposite is true, as for instance in the case of climate change discussed above.

actually get implemented; what will result from the cause and from its method of implementation, where both are subject to the action of the other causes and interferences that will occur; and not just what happens with respect to the effect in question – we need to know about harmful and beneficial side effects as well. So we need more than piecemeal knowledge of what causes what. We need a causal model.[35] Again, our methodologies for how to construct causal models for new target situations from even highly stable well-warranted causal claims are very poor.

3.4   Conclusion

We do have good methods for warranting knowledge claims in the social sciences. The more secure they make the conclusion, though, the more background knowledge we must have in order to apply them. So social science is hard, but not impossible. Nor should that be surprising; natural science is exceedingly hard and it does not confront so many problems as social science – problems of complexity, of reflexivity, of lack of control. Moreover the natural sciences more or less choose the problems they will solve but the social sciences are asked to solve the problems that policy throws up. And here I think we do find special problems for social science. We have very good methods for gathering social science knowledge but considerably less good advice about how to put it to use. So, I urge, what we most need to study, is not how to do social science but how to use it.

[35] See ch. 16 in this book as well as Reiss and Cartwright (2004).

4   Where is the theory in our 'theories' of causality?

4.1   Introduction

Causality is a hot topic today both in philosophy and in economics; there are approaching a dozen different theories of causality on offer. Few are reductionist – they do not embrace the Hume programme of replacing causation by something weaker. But all claim to be 'informative' theories, to tell us what are the central 'characterizing' features of causation. Here is a list of just some of these theories and some of their major proponents. (I focus on theories of generic-level causation, or of 'causal law'. I choose this particular list because these are all theories that are closely related to practice and that I have studied fairly closely.) It certainly does not include all the accounts available.

'Theories' of causality:[1]
- the probabilistic theory of causality (Patrick Suppes) and its descendants
    - Bayes-nets theories (Wolfgang Spohn, Judea Pearl, Clark Glymour);
    - Granger causality (economist Clive Granger);
- modularity accounts (Pearl, James Woodward, economist Stephen LeRoy);
- manipulation accounts (Peter Menzies, Huw Price);
- invariance accounts (Woodward, economist/philosopher Kevin Hoover, economist David Hendry);
- natural experiments (Herbert Simon, economist James Hamilton);
- causal process theories (Wesley Salmon, Philip Dowe);
- the efficacy account (Kevin Hoover);
- counterfactual accounts (David Lewis, Hendry, social scientists Paul Holland and Donald Rubin).

What is my worry about the theory in these theories of causality? Rather than explaining the worry directly as I now see it, I shall instead describe a web

This chapter first appeared in the *Journal of Philosophy*. It was also presented at the 2005 annual conference of the British Society for the Philosophy of Science and at Columbia University. I would like to thank both audiences for helpful discussions.
[1] For more on these see Suppes (1970); Spohn (2001); Pearl (2000); Spirtes, Glymour and Scheines (1993); Woodward (2003); LeRoy (2004); Hoover (2001); Simon (1953); Hamilton (1997); Cartwright (1989); Salmon (1984); Dowe (2000); Lewis (1970); Hendry (2004); Price (1991); Menzies and Price (1993); Granger (1980); Holland and Rubin (1988).

of thought that brought me to it. In the end I do not think one needs to share many of the views I will describe in order to share my concerns. But I think that working through them with me will make the problem more vivid and provide it with a kind of texture and depth that might otherwise be missing.

There are seven interwoven strands that make up this web of thought:

1 causation: one word, many things;
2 causality: metaphysics and methods;
3 representation: handsome is as handsome does;
4 causal laws and effective strategies;
5 hunting causes versus using them;
6 causality: metaphysics, methods and use;
7 where is the theory in our 'theories' of causality?

4.2   On 'theories' of causality

4.2.1   Causation: one word, many things

Generally when I open my eyes and look around me, I see a dappled world, plotted and pieced, not one homogeneous sweeping plain. This tendency has been reinforced by my studies of causality over the last few years, which lead me to the conclusion that causation is not one monolithic concept; nor is there one single thing – the 'causal relation' – that underpins our correct uses of that concept. There are a variety of different kinds of relations picked out by the abstract term 'causes' and a variety of different – correct – uses of the term for a variety of different purposes, with little of substantial content in common.

The variety of theories of causal law on offer provides one of the major reasons in favour of this plurality view. Each seems to be a good treatment of the paradigms that are used to illustrate it, but each has counterexamples and problems. Generally the counterexamples are provided by paradigms from some one or another of the other theories of causality. Usually in these cases we can see roughly why the one treatment is good for the one kind of example but not for the other, and the reverse. Sometimes we do better: we can give an explicit characterization of the kind of system of causal laws that the treatment presupposes and in the nicest cases we can prove a kind of representation theorem. (This is what I think we should always be aiming for.) Bayes nets are a good case. A Bayes-net account of what causality is, of the kind Wolfgang Spohn offers,[2] is provably correct for a system of causal laws over a causally sufficient set of quantities[3] for which the three Bayes-nets postulates (causal

---

[2] See Spohn (2001).

[3] Roughly, this is a set of quantities such that 'all' common causes of quantities in the set are themselves in the set. The scare quotes around 'all' indicate that this needs a far more careful statement since all causes of common causes are themselves common causes and we may not want a condition that strong.

Markov, minimality and faithfulness) hold plus the assumption that if any two factors are probabilistically dependent then either one of them causes the other or they have a common cause. Similarly I prove a representation theorem that shows that for systems of laws that satisfy the axioms of (what I call) a 'linear deterministic system', Woodward's (level-) invariance account of causation always yields correct results.[4]

If this account is correct then we have a variety of different kinds of causal system and what we indicate by saying that one quantity causes another varies according to what characterizes the different systems. This makes a problem for method.

4.2.2   Causality: metaphysics and methods

Metaphysics and method should march hand in hand. If we have an account of what causation is then we ought to be able to explain why our methods for licensing causal claims are good methods for finding just the thing we say causality is. If our account of causation does not mesh with our best methods for finding out about causes, something has gone wrong on one side or the other.

We are used to thinking of causation as one thing, with many different methods to learn about it. Life is more complicated if we accept that there are different kinds of causation with different characterizing features. We must then face up to the question of which methods are appropriate for which kinds of causation. In fact for many of our philosophic theories of causality the question is not so difficult to answer since the theories themselves are almost read off from one method or another. Consider Suppes's probabilistic theory of causality,[5] which also forms the basis for Bayes-nets theories.

Probabilistic 'theory' of causation   Theory:

$$C \text{ causes } E \text{ in } K_i \text{ iff } P(E/C\&K_i) > P(E/\neg C\&K_i)$$

(K_i is a state description over a 'complete'[6] set of confounding factors).

Method: '*C* causes *E*' is licensed if C and E are probabilistically dependent once we have stratified on a 'complete' set of confounding factors.

For varieties of causation where methods are not so immediately apparent, clearly it is a pressing matter to find appropriate methods and to show that they

---

[4] For this interpretation of the theorem in ch. 10, see Cartwright (forthcoming).
[5] Suppes (1970).
[6] The scare quotes are around 'complete' because it is a difficult notion to define. In my opinion it can only be defined relative to a particular choice of causal structure. That is why it has been so difficult to get an exact formulation for the general case (see for instance all the difficulties laid out in Cartwright (1989)).

are appropriate. This was a project that a number of us worked on for several years, sponsored by the Arts and Humanities Research Board.[7]

4.2.3   Representation: handsome is as handsome does

I argue for (borrowing an expression of Maria Carla Galavotti[8]) plurality in causality. But I still look for something in common. Is there not something by virtue of which it is correct to call all these different kinds of relations 'causal'? Here is a proposal.[9] Rather than looking for one special relation in the world that legitimates representing them as causal, look instead for some unified features of the representations themselves. Clearly under this proposal we would not want to be thinking in terms of some kind of 'Fido'-Fido or correspondence theory of representation.

So I turn instead to a theory of scientific representation championed by Mauricio Suarez[10] and Chris Swoyer.[11] A scientific representation should allow us to make inferences about the system represented. Whether or not this is a good starting point for thinking of scientific representation in general, it might be a help in thinking about what makes it right to call a variety of different kinds of relations all 'causal'. The reason for being optimistic about this suggestion is that there is a certain kind of inference that has always been stressed as central to the notion of causation, to which I turn next.

4.2.4   Causal laws and effective strategies

The idea that causes allow us to affect the world has always been part of our thinking about the notion of cause: it is at the core of our current manipulation theories as well as some invariance theories, it is probably the dominant idea about causation in economics right now and it was right at the fore when talk of causation first reappeared, defying positivist prohibition, as witnessed by the title, 'Causal laws and effective strategies', of my first paper on causality.[12] The idea suggested by this in combination with the inference-based approach to representation is that representing a set of relations as causal should allow us to make some kinds of inferences that allow us to use causes as strategies for producing effects. Is something like this possible?

Let me start with some theories of causality where such inferences hold centre place. I shall illustrate with two theories coming from economists rather than any

---

[7] Arts and Humanities Research Board-funded 'Causality: metaphysics and methods', CPNSS, LSE, 2002–4. For more information see http://www.lse.ac.uk/collections/CPNSS/projects/ConcludedProjects/causalityMetaphysicsAndMethods.htm.
[8] Galavotti (2005).   [9] This proposal is also discussed in this book, ch. 2.
[10] Suarez (1999).   [11] Swoyer (1991).   [12] Cartwright (1979).

of the manipulation theories of philosophers for reasons that should be apparent later. Kevin Hoover in *Causality in Macroeconomics*[13] distinguishes between quantities we control directly and those we do not. His aim is to characterize the causal relations between the latter using the concept of 'direct control'.[14] He then takes the following as the central characterizing feature of causality:

x causes *y* iff anything we can do to fix *x* affects *y* but not the reverse.

If that is what causality is, then the inference from '*x* causes *y*' to 'we can influence *y* by fixing *x*' is built right into the metaphysics.

Similarly for econometrician David Hendry:[15]

– Causes are superexogenous.
– Superexogeneity = weak exogeneity + 'invariance'

Weak Exogeneity:
Given $P(y \& x, \beta U \gamma) = P(y|x, \beta)P(x, \gamma)$,
x is exogenous to a vector of outcome variables *y* if the parameters γ of the marginal distribution of *x* have no cross-restraints with the parameters β of the conditional distribution of *y* on *x*.[16]

Superexogeneity:
In addition the conditional distribution is invariant to interventions that change (parameters of) the marginal distribution.

Suppose then that *x* is weakly exogenous with respect to *y* and we think of changing the distribution of *x* in order to affect the distribution of *y*. It may seem that we can predict the outcome from the formula for the conditional distribution. But that is not so. Changing γ changes the joint probability distribution and there is nothing that ensures that the new conditional distribution will still be the same. In the original distribution β and γ may have no dependencies but that does not show what happens if the distribution is changed. So Hendry adds the constraint: *x* causes *y* only if the parameters, β, of the conditional distribution, $P(y|x)$, stay fixed as we vary the parameters, γ, of the distribution of *x*. Again an inference about use follows immediately from the very notion of causation.

Can we do this in general? Recall, that was my hope: representing a relation as causal allows some kinds of inferences about use. Unfortunately this hope is not supported by the other theories of causality on offer. To see this vividly consider James Woodward's manipulation/invariance theory of causality, a case

---

[13] Hoover (2001).   [14] For more about Hoover's proposal see ch. 14 of this book.
[15] See Engle et al. (1983) and Hendry (2004) plus the discussions of Hendry in ch. 16 of this book.
[16] This is only a special case of the definition, for illustration. Weak exogeneity implies that the marginal distribution can be ignored in estimating the conditional distribution.

where we might expect the connection to be immediate just as in the Hoover and Hendry theories.

4.2.5   *Hunting causes versus using them*

It is often supposed[17] that the 'gold standard' for hunting causes – i.e. for ascertaining causal relations – is the ideal Galilean experiment. In this kind of experiment we vary the cause holding fixed 'all' other sources of variation of the putative effect and look to see if the effect varies in train. This is just what James Woodward uses to characterize causation.[18] He considers situations in which the putative effect, y, is a function of the putative cause, x; so they co-vary. If x is to be correctly said to cause y then he demands the very strong condition that x be able to vary independently of all other sources of variation in y as we should like to see in a Galilean experiment. This condition is called modularity. He also requires that the functional relation between x and y stay the same across variations in x that keep fixed the other causes of y. This ensures that y will co-vary with x in a Galilean experiment, as required if x is to be judged a cause of y from the experiment (but it is stronger than required to ensure the covariation). This condition is called invariance. So . . .

Given $y = f(x, \ldots)$, x causes y iff
1 modularity: x can be varied independently of all 'other' sources of variation in y
2 invariance: $y = f(x, \ldots)$ continues to hold as x is varied holding fixed other sources of variation in y

This might reasonably be called a manipulation account but it is very different from the two manipulation accounts from economics that I have just reviewed. Woodward's theory of causality licenses counterfactuals about what would happen if the cause were to vary in a very special way, a way that is not what we would normally be envisaging for either policy or technology. The variation is just the kind we need if we want to test a causal claim – to hunt for causes – but not the kind we expect to implement when we try to use them. The same is true for many of the other accounts. Recall that many are almost read off some one or another method for establishing causal claims. Correlatively, like Woodward's account they give little guidance about what we can do with a causal claim once we know we are entitled to it.

This observation is true almost universally across our list of causal theories. They are in essence either 'language-entry rules' – they tell us when it is correct to label relations 'causal' (and generally only one method for so doing) – or

---

[17] Wrongly I think. See chs. 3, 7–10 and 16 in this book.
[18] This is my interpretative summary of what Woodward's account amounts to in brief and not literally how he defines things. See Woodward (2003) and my discussion of his invariance views in chs. 6–10 in this book.

they are 'language-exit rules' that tell us what inferences we can draw from that label (and again, usually only one particular kind of inference). This raises the twin concerns:

• What is the use of causes characterized from a hunting methodology?
• How do we hunt causes when they are characterized in terms of their use?

4.2.6   Causality: metaphysics, methods and use

I have briefly sketched the first of my twin problems in referring to Woodward's manipulation/invariance account. The same holds for Lewis-style counterfactual accounts. They provide an elaborate procedure for deciding when we can attach the label 'cause'. But then what? There is nothing more in the account that allows us to move anywhere from that, nothing that licenses any inferences for use.

The second concern can be readily seen in Milton Friedman's famous arguments in favour of hugely idealized models in economics.[19] Models, he urges, should be judged by the correctness of their results in the domain under investigation, not by the correctness of their assumptions. So like the economists' accounts of causality I discussed above, for Friedman models are geared to use. But this very fact restricts their usefulness. The most useful models are ones we can first legitimate by some independent means and then use to make predictions about new matters. On Friedman's account all there is to the acceptability of the model is that it gives correct predictions about the matters we are interested in – and that is just what we cannot know about the model beforehand if our point in using it is to gain confidence in those predictions. Similarly for Hoover's account of causality. What we should like is to have an independent reason for thinking that x causes y, then use that to infer that manipulating x will affect y. But for Hoover that is just what it is for x to cause y so we cannot establish the causal relation first and then use it later for policy predictions.

What we see from the twin concerns above is that in all the 'theories' of causation listed, the metaphysics is too thin. It is geared either to hunting causes or to using them but not to both. My own earlier project, *Causality: Metaphysics and Methods*, missed this point. Yes, the metaphysics of causality must be tied to methods on the one hand, but it must equally be tied to use on the other. As I urge in the introduction to this book:

Our philosophical treatment of causation must make clear why the methods we use for testing causal claims provide good warrant for the uses to which we put those claims.

I should stress that this same requirement holds equally for a universalist and for a pluralist view of causality. Our notions of causation, or our characterizations

[19] Friedman (1953).

of specific kinds of causal systems, must be rich enough to allow us both to ascribe the label 'causal' and then to make some use of it once we have done so.

4.2.7 Where is the theory in our 'theories' of causality?

Considering Hendry in this light raises yet another issue. Weak exogeneity is a concept having to do with statistical inference – estimating what a distribution is. This has little apparent connection with causality. But Hendry's attention to the conditional distribution of y on x – $P(y/x)$ – in his account of 'x causes y' is reminiscent of the probabilistic theory of causality. Then he adds on top an invariance condition. It is not that we do not have available any methods for testing for the kind of invariance that Hendry demands. Econometrics has good methods for looking for 'structural breaks' – places where a distribution changes. The idea then is that we look to see if structural breaks in the distribution we will use for prediction are associated with changes in the parameter we propose to manipulate for policy.

What is disappointing from the point of view of the metaphysics of causality is that the invariance condition is, as I said, added on. And it is what does the work of licensing predictions for use, not any special features of the thing that is itself invariant. Any kind of formula that allows us to predict what we want to know and that holds true across the variations we envisage making will do; it need not have any of the other kinds of characteristic that are in the cluster we usually associate with causality – increase in probability of the effect on the cause, space–time connection, asymmetry of the cause–effect relation, existence of a mechanism, etc.[20]

This is the point that Sandra Mitchell makes about the notion of 'scientific law'.[21] For use in prediction we do not need a claim that holds universally; nor, it seems, do we need anything that satisfies any of our usual accounts of what it is to be a 'causal' law. We only need a claim that holds across the range of situations in which we will use it for prediction. Sometimes the same point seems to be urged by Woodward too. He seems to argue that invariance – and just invariance – across the required, possibly highly limited, range of envisaged uses is what matters. We might call that 'causation' then. But he does

---

[20] We can see a good instance of this in my discussion of Hoover's theory of causality in ch. 14. There I argue that Hoover's 'strategy' relations are distinct from 'production' relations, which have many of the other features generally associated with causality. On the other hand, the kinds of production relations described there can have just the drawback I am worrying about here – nothing seems to follow from them about strategy, except the truism that if the equations continue to be accurate across the changes we will make, they can be used for prediction of what would happen were we to make those changes.

[21] Mitchell (2003).

not embrace this: he insists on the Galilean manipulation account of causation that I described, which is not geared to practical use.

Bayes-nets methods make some advance here. I think for instance of the manipulation theorem of Spirtes, Glymour and Scheines,[22] that computes facts about the joint probability of sets of variables after manipulation from facts about their distribution before plus facts recorded in the causal graph, or of Pearl's thick book on counterfactuals.[23] I worry though that both cases are still too close to the strategy I ascribed to Hendry – the 'add on' strategy.

Bayes nets characterize causality in terms of the three axioms I cited, the chief of which is the causal Markov condition. But how important is the causal Markov condition to our ability to predict facts about the probability distribution after manipulation from knowledge of the causal relations plus the probability distribution before? I do not think it can be too important.

Consider for example what in ch. 10 of this book I call 'epistemically convenient systems' with probability measures. For these we can define a very close analogue of the Spirtes–Glymour–Scheines concept of manipulation, including, as they do, the assumption that the causal relations are invariant under manipulation. Systems like this need not satisfy the causal Markov condition. Yet we can derive results about what happens to the joint distributions under manipulation in these kinds of system just as we can when the causal Markov condition is satisfied.[24] In both cases most of the work in deriving the relevant results is done not by the axioms of the system but by the very restricted notion of 'manipulation' that is used. The manipulations involved in these theorems are like Woodward's (or those miracle manipulations of David Lewis) – the kinds of manipulation we should like to make in a Galilean experiment, not the kind we envisage for policy, where many things may be expected to change at once, including some of the causal principles by which the system operates.[25] The situation is similar for Pearl on counterfactuals: most of the work in legitimating policy predictions comes from the semantics he offers for counterfactuals, not from the special modularity conception of causality he urges (though the two are nicely meshed).[26]

More generally, all the cases that I know where we have counterfactual inferences or manipulation theorems that can be proved from a particular kind of system of causal principles work like this. First, a set of relations are postulated

---

[22] Spirtes et al. (1993).   [23] Pearl (2000).

[24] What the causal Markov condition does is to give the joint probabilities both before and after a particularly nice form. Roughly, conditioning on the direct causes of x, y becomes independent of x if x does not cause y, which means that we compute the probability of y without regard to x.

[25] Cf. the famous Lucas critique of macroeconomics, which depends on the assumption that manipulating policy variables produces changes in the causal principles governing the system. See Lucas (1976).

[26] And, like Woodward and Lewis, that semantics is geared to hunting causes, not using them.

that satisfy a particular set of constraints taken to be sufficient to characterize them as 'causal relations'. Second, manipulations are taken to create a new, closely related system. The new systems are defined by essentially two clauses. The first describes what is to change, e.g. the value of certain variables, maybe representing quantities all of which can vary independently of each other (i.e. they are 'variation free'), or the relation by which the 'manipulated' quantity is produced. The second states a set of things that do not change, like values of other particular variables and all the other specified relations except that by which the 'manipulated' quantity is produced. Then the counterfactual consequence is calculated in the new system.

What is striking about this is that the constraints that justify labelling the system of relations as 'causal' seem to have no role to play. What matters is that there is a system of relations, say equations, that allow us to solve for a given result. Then we are given a recipe for making very specific changes and the recipe produces a new system that also allows us to solve for the result in question. The solvability of the two systems plus the rule for getting from one to the other is what does the job, not any of the constraints that make it a 'causal' system in the first place (beyond those that ensure solubility). So what in the end does causation have to do with policy counterfactuals, manipulation or effective strategies?

I have been hoping for something more, something that shows why our methods should warrant causal claims with practical strategy implications – and something of substance that does so. It is not enough for a metaphysics of causality to adopt just the 'add-on' strategy, to be simply a union of a language-entry rule and a language-exit rule. What we should be looking for is a theory of causality, in much the same way as we have a theory of the electron. The account itself should combine with other chunks of knowledge to imply both the language-entry and the language-exit rules.

4.3   Conclusion

When it comes to theories of causal law, I have earlier urged that we have on offer not a plurality of competing theories but rather a host of theories each appropriate to one of a large plurality of different kinds of (systems of) causal laws. I now worry that this was too generous. Rather than a plurality of theories, we do not have any theories at all.

I close with a hypothesis to think about. Bertrand Russell maintained that there are no causal claims in physics.[27] Many of us have urged the contrary. Physics is rife with causal claims, for instance Einstein and Infeld: 'Forces cause motions: Gravitational forces, along the line of action; electromagnetic,

---

[27] Russell (1913).

parallel to the line of action.' If so then what we have in each of the specific domains of physics are theories about that domain, theories that contain causal claims. 'Massive bodies attract each other.' Obviously the same could be said across the sciences. Perhaps we should find our lead here. Maybe our whole enterprise is misplaced. Perhaps we should be seeking not theories of causality but rather causal theories of the world.

Case studies: Bayes nets and invariance theories

5   Preamble

Part II looks in detail at two different kinds of theories of causality, Bayes-nets theories and various invariance or 'modularity' theories. I focus on these principally out of personal history. Both are closely associated with probabilistic theories of causality, which I have worked on in the past. In keeping with the pluralistic stance of this book, I take that it these are not best seen as alternative accounts of one kind of relation – the causal relation – but rather as descriptions of different kinds of systems of different kinds of causal relations requiring different methods to learn about them.

Both kinds of account are closely related to method. Bayes-nets methods provide a way to discover the different causal arrangements consistent with given input information, especially information about conditional independencies and any known causal relations. And at least one version of the invariance accounts – the one I discuss in ch. 10 – can be immediately translated into a method for testing for causality. What is nice in both cases is that it is possible to provide an explicit characterization of the kinds of systems to which the account applies and then to prove that the methods are appropriate to those kinds of systems. For Bayes nets this kind of proof has been there from the start. The account is presented axiomatically, then the methods are derived from it. The proofs in ch. 10 do the same kind of job for certain kinds of invariance. One contribution of this chapter that I feel bears special note is its treatment of regression equations in the second theorem. It is common in both philosophy and economics to take regression equations as one way to represent causal principles, yet there is little discussion of what it means for these to be a correct representation.[1] Chapter 10 provides an explicit characterization.

Although not explicitly cast in this form, the two theorems of ch. 10 are a kind of representation theorem.[2] A causal system is characterized explicitly by a set of axioms; then it is proved that invariance (as formally characterized) is an adequate representation of the relations in the system; that is, given the

---

[1] James Woodward is an exception. Consistent with his view that invariance is what causality consists in, he takes them to be correct if they satisfy certain invariance requirements. As I argue in chs.7 and 8 I do not share his view that invariance is essential to the notion of causality.

[2] For a presentation of an invariance theorem in this form see Cartwright (forthcoming).

characterizing features of the causal system, a hypothesized relation is causally correct if and only if it is invariant.

As I urged in the introduction, if something is to be accepted as a test for a given kind of causality, we should be able to prove that it is a good test. James Woodward, who has championed the kind of invariance account under study in ch. 10 asks how we could do that. Invariance is as basic as it gets he thinks. For Woodward, invariance (of the right sort) is the characterizing feature of causality.

The axioms of ch. 10 lay out what is more basic in this case. A principal one is the axiom of causal priority requiring that all functionally true relations be the result of the fundamental 'causal' relations. Woodward himself relies on just this fact in his less formal defences of invariance.[3] But it is not one that can be assumed to hold in every system. Genuine functional relations need not arise from an underlying causal structure but may hold because of boundary

---

[3] The other axioms, with one possible exception, are all ones that it seems Woodward accepts as well. The exception is the axiom of numerical transitivity. Woodward rejects the idea that causal relations are transitive. But his arguments do not seem to have any bearing on the specific kind of transitivity assumed in this axiom. For instance, he supports Hitchcock's (2001) claims that causality is not transitive. But Hitchcock's treatment there supposes the same kind of 'numerical transitivity' that is assumed in ch. 10. To appreciate the thinness of the connection between conventional philosophical worries about transitivity and numerical transitivity, consider a stock example. Jack wants to explode a bomb if he gets it wired up. (Let $W = 1$ for getting the bomb wired; $W = 0$ otherwise.) If he does so he will push the button with his thumb. (Let $P = 1$ for pushing the button with his right thumb; $P = -1$ for pushing it with his left; $P = 0$ for not pushing it.) A dog bites his right thumb. (Let $D = 1$ for dog bites right thumb; $D = 0$ for dog does not bite at all.) So Jack pushes the button with his left thumb and the bomb explodes. (Let $X = 1$ for bomb explodes; $X = 0$ for bomb does not explode.) The dog's bite causes Jack to push the button with his left thumb and pushing the button with his left thumb causes the bomb to explode. But we may be reluctant to allow that the dog's bite caused the bomb to explode. If so, there is a failure of transitivity. Presumably a good part of the reason for this is that it seems the bite only influences the way in which the explosion is brought about, not whether it occurs; the explosion will occur whether the bite occurs or not.

Suppose now we think of this as a repeatable situation with the principles governing it as described. Here is a simple way to represent them (though it is not linear):

$$P = W - 2WD \tag{1}$$
$$B = P^2 \tag{2}$$

By numerical transitivity

$$B = W^2 - 4W^2 D + 4W^2 D^2 \tag{3}$$

Equation (3) not surprisingly gives the correct results and it is also invariant to interventions over the variable set $\{W, D\}$, which means that Woodward's own criterion judges it causally correct relative to the set $\{W, D, P\}$ containing the 'later' variable P. This shows that it does not represent the most direct causes relative to the expanded variable set, which does not counter the fact that it is causally correct relative to $\{W, D\}$. Moreover equation (3) itself shows why we may be reluctant to say that the occurrences of the dog bite – the event of D taking value 1 – causes the occurrence of the explosion – the event of B taking value 1. For according to equation (3), so long as the bomb is wired ($W = 1$), it explodes ($B = 1$) whether the dog bite occurs ($D = 1$) or not ($D = 0$).

conditions, conservation laws, imposed symmetries, shared time trends and for various other reasons.

The other chapters in part II make similar cautions about limitations of scope. Causal priority can justify the use of certain invariance methods but it is not a universal characteristic of causal systems. Bayes-nets methods are powerful – but only for systems that satisfy the Bayes-nets axioms. The kind of invariance that gets labelled 'modularity' is a fine feature for a system of relations to have and it meshes nicely with Judea Pearl's semantics for counterfactuals discussed in ch. 16. But it does not characterize every system we would like to label 'causal'.

There are two chapters here – chs. 8 and 9 – on the connection between modularity and Bayes nets. They concern attempts to prove that modularity implies the causal Markov condition, which is so central to Bayes-nets methods. I show that the proofs on offer are invalid and explain why no connection should be expected. The issues are in one sense very local, looking at the details of specific proofs. But the fact that this connection, so strenuously defended by other researchers, fails does back up my claims that modularity and Bayes-nets methods are appropriate to different kinds of causal system.

The two kinds of accounts of causality discussed here are only a selection from the list compiled in ch. 2 and that list is surely not complete. But these two provide a model for how I believe metaphysics and methods should be linked in every case. For each metaphysical account there should be an explicit characterization of the kinds of system to which it applies; then a proof that the account and the methods it spawns are appropriate to that kind of system. That is, we need representation theorems for all our theories of causality. Only then can we be assured that the methods we use can warrant the conclusions we draw from them.

The invariance account of ch. 10 and Bayes nets also have a second exemplary feature. Recall that at the conclusion of ch. 4 I suggested that perhaps we do not need theories of causality so much as we need causal theories of the world. My original idea is the one that introduces this book: we need to look for richer theories of what causality is that will show us both how to test for it and how to use it. The alternative suggested in ch. 4 is that we might make more progress by studying specific scientific theories that describe specific causal relations. The methodological contribution would be to get clear in each case how those content-rich causal claims can get tested and how they can be used.

The Bayes-nets and invariance accounts of ch. 10 stand neatly between these two projects. They provide templates of forms that specific content-rich causal theories can take. Consider the invariance account. It is formulated for what I label 'linear deterministic systems'; this gives an abstract schema that various content-rich systems might fit. Similarly for the directed acyclic graphs of Bayes nets. When systems do fit these templates, we have ready-made methods for discovery or testing as well as ready-made theorems for use. Can the other 'theories' of causality on offer provide us with similarly fertile templates?

Despite their advantages, there are outstanding problems for these two schemes. The first I should like to stress concerns the central theme of this book. Both are far better geared to hunting causes than to using them. Bayes nets are the better off in this respect than invariance methods. There are both manipulation theorems like that of Spirtes, Glymour and Scheines and the rich, detailed treatment of counterfactuals developed by Judea Pearl in tandem with his work on Bayes nets. But for both, the inferences are far more limited in scope than we might hope. I describe how in chs. 4 and 16.

The second disadvantage I should like to underline is closely related. We have little idea how to combine the information encoded in the templates with other information from the more concrete theories that fit the template. After all, the inferences that can be drawn from various kinds of causal claim need not depend just on the logic of those claims but on the network of other claims in which they are embedded. Where we can integrate the two, we can expect to generate far richer sets of inferences that can be put to use for policy and planning.

6   What is wrong with Bayes nets?

6.1   The basic question: can we get to causality via Bayes nets?

Probability is a guide to life partly because it is a guide to causality. Work over the last two decades using Bayes nets supposes that probability is a very sure guide to causality. I think not, and I shall argue that here. Almost all the objections I list are well known. But I have come to see them in a different light by reflecting again on the original work in this area by Wolfgang Spohn and his recent defence of it in a paper titled 'Bayesian Nets Are All There Is to Causality'.[1]

Bayes nets are directed acyclic graphs that represent probabilistic independencies among an ordered set of variables. The parents of a variable X are the minimal set of predecessors that render X independent of all its other predecessors. If the variables are temporally (or causally) ordered, we can read the very same graph as a graph of the (generic-level) causal relations among the quantities represented, it is maintained. This commits us to the causal Markov condition described below, which is a relative of Reichenbach's claim that conditioning on common causes will render joint effects independent of one another. It is also usual to add an assumption called faithfulness or stability as well as to assume that all underlying systems of causal laws are deterministic (plus the causal minimality condition, which I will not discuss). With these assumptions in hand there are a variety of algorithms for inferring causal relations from independencies. These I will loosely call 'Bayes-nets methods'.

In criticizing the inference to causes from Bayes nets it is usual to list the objections I note. Is this just an arbitrary list? And why should one have expected any connection between the two to begin with? After all, Bayes nets encode information about probabilistic independencies. Causality, if it has any connection with probability, would seem to be related to probabilistic dependence.

Research for this paper was supported by a grant from the Latsis Foundation, for which I am extremely grateful. The work is part of the Measurement in Physics and Economics Research Project at LSE.
[1]   Spohn (2001).

The answers to the two questions are related. When we see why there might be a connection between causality and independence, we see why there will be a list of objections. The answer to 'why' is not one that will sound surprising, but I want to focus on it because working through the argument will show that we have been looking at probability and causality in the wrong way. Probabilities may be a guide to causes, but they are, I shall argue, like symptoms of a disease: there is no general formula to get from the symptom to the disease.

6.2   The call for explanation

It is usual to suppose that once the right set of assumptions are made about the causal systems under study, we can read information about causes from a Bayes net that satisfies those assumptions. Wolfgang Spohn maintains that if there is a tight connection like this between Bayes nets and an independent notion of causation, there should be a general reason for this. He cannot find one; so he proposes that the notion of causation at stake is not independent. The probabilistic patterns of a Bayes net *is* our concept of causation: 'It is the structure of suitably refined Bayesian nets which decides about how the causal dependencies run.'[2]

I agree with Spohn that if there is a tight connection, there should be a reason for it. The alternative is what Gerd Buchdahl called a 'brute-force' connection, one which holds in nature but has no 'deeper' explanation. There are such brute-force connections between concepts we use in science. These are what we record in fundamental laws. And some of them involve relational concepts like 'causes'. For instance, if the allowed energy configuration for a system in relation to its environment is represented by a specific Hamiltonian, say H, then whatever the system's current state (say Φ), the system will evolve in time according to $-i/h\partial\Phi/\partial t = H\Phi$.

But I do not think there is a brute-force connection in the case of Bayes nets and causation. That is primarily because there is a reason for the connection, a good reason. The problem is that the reason does not justify a tight connection. The reason lets us see why the connection will hold when it does, but it also allows us to see how loose the connection between the two is. For simplicity I will stick to yes–no causes and effects in the subsequent discussion. We are looking for an equivalence between causal connections and Bayes nets. I will start with causation and see first how – and when – we can get from causation to the probability relations pictured in Bayes nets.

---

[2] Spohn (2001), p. 166.

6.3 From causation to probabilistic dependence

6.3.1 *Where have all the caveats gone?*

Causes produce their effects; they make them happen.[3] So, in the right kind of population we can expect that there will be a higher frequency of the effect (*E*) when the cause (*C*) is present than when it is absent; and conversely for preventatives. What kind of populations are 'the right kinds'? Populations in which the requisite causal process sometimes operates unimpeded and its doing so is not correlated with other processes that mask the increase in probability, such as the presence of a process preventing the effect or the absence of another positive process.

Here are some of the conditions that we know need to be satisfied: the necessary triggering and helping causes for *C* must operate together sometimes in the population, and their joint operation should not be probabilistically dependent on that of other causes or preventatives, or of *C* itself. Nor is the operation of *C* to produce *E* probabilistically dependent on the operation of other causes or preventatives of *E*.

The trouble with Bayes nets is that they ignore all these caveats. When Bayes nets are used as causal graphs, effects are probabilistically dependent upon each of their causes. That's it. Nothing can mask this. The assumptions about causality made by the Bayes-net approach go all the way back to the first, then ground-breaking, probabilistic analysis of causality by Patrick Suppes.[4] Suppes begins with prima facie causation: any earlier factor that is correlated with an effect is a prima facie cause of that effect. Real causes are ones that survive the same independence tests that are required in the Bayes net. But nothing gets to be a candidate for a cause unless it is correlated with the putative effect.

Twenty to thirty years later we have the Bayes-net approach, in all essentials equivalent to the original formulation proposed by Suppes when the subject first began. It is as if Simpson's paradox and causal decision theory never existed. Nor decades of practice by econometricians and other social scientists, who plot not simple regressions but partial regressions. Nor the widely deployed definition proposed by Granger in 1969, which looks for probabilistic dependence only after conditioning on the entire past history of the cause – which ensures that all the other causes up to the time of the putative cause will be held fixed.

The demand that effects always be probabilistically dependent on each cause follows in the Bayes-net approach from the assumption that Peter Spirtes, Clark Glymour and Richard Scheines call faithfulness.[5] Judea Pearl calls it stability.[6] The assumption is necessary to the approach. Without it the procedures

---

[3] When we deal with quantities of more than two values, there are other possibilities, e.g. a cause may raise the level of the effect.
[4] Suppes (1970).   [5] Spirtes, Glymour and Scheines (1993).   [6] Pearl (2000).

developed by Pearl and by Spirtes, et al. cannot get very far with the discovery of causal connections; and the proofs that assure us that they will not make mistakes if sample sizes are large enough will not go through. For Spohn it matters because he argues that causal connections are the connections marked out on God's big Bayes net once the variables have been temporally ordered. With the faithfulness/stability assumption, the causal connections are unique; but they are seldom unique otherwise.

For those readers who are not already deeply into this discussion let me rehearse the standard objections to the assumption that all genuine causes are prima facie causes. First there is Simpson's paradox: facts about probabilistic dependency can be reversed in moving from populations to subpopulations. For example, factor X may be positively dependent or negatively dependent or independent of Y in a population but still be any of these three in all partitions of the population along the values of a third factor Z if Z is itself probabilistically dependent on X and Y. Z may for instance be a preventative of Y; because of its correlation with X, the presence of X does not after all increase the frequency of Y's in the population as my opening argument suggests.

It is typical in social science to sidestep this problem by looking for probabilistic dependence between a putative cause and its effect only in subpopulations in which all possible confounding factors are held fixed. This is apparent in the econometrics concept of Granger causality, where X Granger causes Y if X and Y are probabilistically dependent holding fixed everything that has occurred up to the time of the putative cause.

The same strategy was at the heart of various versions of causal decision theory two decades ago. What is the probability that if I were to do C, E would occur? The conditional probability $P(E/C)$ gives the wrong answer. It could be either way too big or way too small because of the operation of confounding factors; what is relevant here, it could be zero even though my doing C could have a substantial impact on whether E occurred. One standard proposal (the one I urged)[7] is to set $P(C\square \rightarrow E) = P(E/C\&K)$, where K is a state description over the values of a full set of 'other' causal factors for E that obtain (or will obtain) in the decision situation.[8] Where we do not know the values of the factors in K we should average over all possible values using our best estimate of the probability that they will occur: $\Sigma P(E/C\&K_j)P(K_j)$.

Exactly the same formula has reappeared among the Bayes-nets causal theorists. Judea Pearl has recently produced a very fine and detailed account of

[7] Cartwright (1983), 'Causal Laws and Effective Strategies'.

[8] What counts as a complete set of factors is not so easy to characterize for probabilistic causality. (For one definition, see Cartwright (1989), p. 112.) The task is easy if we are allowed to help ourselves to the notion of the objective system of causal laws governing a population, as Pearl and Spohn and I do (see ch. 10 here). In that case K ranges over all the parents of E, barring C, relative to God's great causal graph for the population.

counterfactuals and their probabilities, based on Bayes nets. According to Pearl, the probability of y if we were to 'set' $X = x_j$ is $\Sigma P(y/x_j$ & parents of $x_j)P$ (parents of x_j).[9] Despite Pearl's endorsement of stability, from this formula it looks as if a factor can have a high degree of causal efficacy even though on his account it is not really a cause at all because it is not prima facie a cause. I take it that Pearl does not take this to be a problem because he thinks cases where 'stability' is violated involve ' "pathological" parameterizations'[10] and are not in the range he will address.

The second kind of case usually cited in which genuine causes are not prima facie causes is when one and the same cause has different kinds of influence on the effect. The different influences may cancel each other. The easiest version of this to handle is when a given factor acts as both cause and preventative of the effect, by different routes. G. Hesslow's birth-control pills[11] are the canonical philosophical example. The pills are a positive cause of thrombosis. On the other hand, they prevent pregnancy, which is itself a cause of thrombosis. Given the right weights for the three processes, the net effect of the pills on the frequency of thrombosis can be zero.

If we suspect that cancellations of this kind are occurring, we can confirm our suspicions by looking at the probabilities of thrombosis given the pills in populations in which factors from the separate causal routes between pills and thrombosis are held fixed. But this is no comfort to the Bayes-nets theorist.

Even this strategy is not available where there are no routes between the cause and its effects. We have two kinds of trouble with routes. The first is a worry that I share with Glenn Shafer.[12] Let us accept that in every case of singular causation there is a temporally continuous process connecting the cause with the effect. That does not guarantee that there will always be a vertex between any two other vertices in God's great causal graph. That is because the graphs are graphs of causal laws that hold between event-types. Every token of the cause-type may in actuality be connected by a continuous process with the effect token without there being some chain of laws in between: C_1 causes $C_2, \ldots, C_n$ causes E. For instance, the signal from the cause to the effect may piggyback on causal processes that the cause in question does not initiate. This is particularly likely where there are a lot of processes with the necessary spatio-temporal relations already available for use in conveying the influence from the cause to the effect.[13]

Roman Frigg offers a number of examples.[14] He explains that what is wanted are cases for which there is:

A generic law without contiguity: on the generic level cause and effect do *not* exhibit contiguity, neither in space nor in time.

[9] Pearl (2000), p. 73.   [10] Pearl (2000), p. 48.   [11] Hesslow (1976).   [12] Shafer (1996).
[13] For further discussion, see Cartwright (1999), chs. 5 and 7.   [14] Frigg (2000).

No unique causal chain: there is no unique causal chain that connects cause and effect. That is, on the *concrete level* the connection between the two can be realised in many different ways.

No unique transmission of causal information: the causal information may be transmitted in different ways. (In addition the kind of physical and/or institutional structures that guarantee the capacity of the cause to bring about its effect may be totally different from those that guarantee that the causal message is transmitted, i.e., the causal law and the individual chains connecting the cause and effect may result from different structures.)

One of Frigg's examples involves as cause person B getting an HIV virus from another person and, as effect, that B dies later on. He tells us:

1 The infection with the HIV virus leads in most cases to death. But a long period of time elapses between these two events . . .
2 The infection with the HIV virus leads (in most cases) to the outbreak of AIDS, i.e. the destruction of the immune system. This in turn can lead to death by a variety of different routes. To mention just a few: diarrhoea (various pathogenes possible); encephalitis with brain atrophy; neuropathy; pneumonia; ringworm (various types); meningitis; herpes simplex; tuberculosis; and fever.
3 The causal information can initially be transmitted on various different paths as well: sexual contact (vaginal, oral, anal), exchange of blood (blood trans-fusions, use of dirty needles, injuries in the hospital), communication from the mother to the child.[15]

Frigg also offers examples of death by malaria and from exposure to strong radiation, and of the democratic election of an individual as president of a country resulting in that person's becoming president, the receipt of a court order causing someone to appear in court and the ordering of a plane ticket causing a person to receive a ticket.

The other is a problem we all sweep under the rug: the representations in a causal graph are discrete; for every vertex there is always a predecessor and a successor. Is causality really like that? If it is then we can have cases of causes with mixed influences on the effect, directly, not by different routes. So the device of holding fixed vertices along the various routes is not even available to provide us with a way of using facts about probabilistic dependencies and independencies to test whether a factor is really causally inefficacious rather than having mixed influence on the effect.

6.3.2   Can the caveats be ignored?

What justification do Bayes-nets theorists give for ignoring all these caveats and insisting that all causes must appear as causes at the first crude look? Spirtes, Glymour and Scheines discuss Simpson's paradox at length. They present two

---

[15] Frigg (2000), p. 1.

graphs, the first embodying Simpson's paradox; the second is a graph that by contrast is 'faithful' to all the independencies assumed in the paradoxical case, i.e. as in Suppes's original formulation, all causes are prima facie causes:[16]

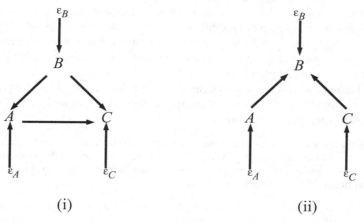

(i)   (ii)

Figure 6.1

They then invite us to

[s]uppose for a moment that we ignore the interpretation that Simpson gave to the variables in his example . . . Were we to find A and C are independent but dependent conditional on B, the Faithfulness Condition requires that if any causal structure obtains, it is structure (ii). Still, structure (i) is logically possible,[17] and if the variables had the significance Simpson gives them we would of course prefer it. But if prior knowledge does not require structure (i), what do we lose by applying the Faithfulness Condition; what, in other words, do we lose by excluding causal structures that are not faithful to the distribution [i.e. that allow genuine causes that do not appear as causes *prima facie*]?[18]

I assume that this passage is meant as a defence of the faithfulness condition since it appears at the end of the long exposition of Simpson's paradox in the section in which they introduce faithfulness as an axiom[19] and just before the only other remark that could be construed as a defence of this axiom in the face of Simpson's paradox. But what is the defence? The answer to their question is obviously: what we lose is getting the causal structure right.

[16] Spirtes et al. (1993), p. 68.
[17] I suppose they mean by 'logically' possible that it is consistent with the other assumptions they wish to make about causal laws and probabilities.
[18] Spirtes et al. (1993), pp. 67–8.
[19] In fact this is not literally true since the section, though headed 'Axioms', only introduces a definition of faithfulness and does not make any claims about it. It is clear from the various sales pitches they make for their methods, however, that they take it to be a condition true of almost all causal systems.

Perhaps they mean to suggest that when we do not know anything, it is more reasonable to plump for structure (ii) than for structure (i). But what is the argument for that? I respond with a truism: when you don't know, you don't know; and it is often dangerous to speculate. If we have no idea what the variables stand for, let alone how they operate, we are not in a position to make a bet with any degree of credibility. 'Ah yes', I am sometimes told, 'but what if you *had* to bet?' Well, tell me more about the context in which I am forced to bet – a psychological experiment perhaps? – and I may be able to tell you which bet I would plump for.

Perhaps, however, Spirtes, Glymour and Scheines are speaking sloppily. They do not mean '*what* do we lose?' but rather '*how often* will we lose?' For immediately after this they report that

[i]n the linear case, the parameter values – values of the linear coefficients and exogenous variances of a structure – form a real space, and the set of points in this space that create vanishing partial correlations not implied by the Markov condition [i.e. that violate faithfulness] have Lebesgue measure zero.[20]

This is surely intended as an argument in favour of faithfulness – and it is frequently cited as being so intended – though I am not sure exactly what the conclusion is that it is supposed to support. I gather we are to conclude that it is unlikely that any causal system to which we consider applying our probabilistic methods will involve genuine causes that are not prima facie causes as well.

But this conclusion would follow only if there were some plausible way to connect a Lebesgue measure over a space of ordered n-tuples of real numbers with the way in which parameters are chosen or arise naturally for the causal systems that we will be studying. I have never seen such a connection proposed; that I think is because there is no possible plausible story to be told. Moreover, were some connection mooted, we should keep in mind that it could not bear directly on the question of how any actual parameter value is chosen because, as we all know, any specific point in the space will have measure zero. So we not only need a story that connects a Lebesgue measure over a space of n-tuples of real numbers with how real parameter values arise, but we need a method that selects as a question to be addressed before values are chosen: shall values occur that satisfy faithfulness or not?

Not only is the theorem about the Lebesgue measure not relevant to the issue of whether all causes are prima facie causes. I think it is an irresponsible interjection into the discussion. Getting it right about the causal structure of a real system in front of us is often a matter of great importance. It is not appropriate to offer the authority of formalism over serious consideration of what are the best assumptions to make about the structure at hand.

---

[20] Spirtes et al. (1993), p. 68.

Judea Pearl argues somewhat differently about the choice of parameter values. He uses the term stability for the condition that insists that effects be probabilistically dependent on their causes even before confounding factors are conditioned on. Here is what he says in its entirety:

Some structures may admit peculiar parameterizations that would render them indistinguishable from many other minimal models that have totally disparate structures. For example, consider a binary variable C that takes the value 1 whenever the outcomes of two fair coins (A and B) are the same and takes the value 0 otherwise. In the trivariate distribution generated by this parameterization, each pair of variables is marginally independent yet is dependent conditioning on the third variable. Such a dependence pattern may in fact be generated by three minimal causal structures, each depicting one of the variables as causally dependent on the other two, but there is no way to decide among the three. In order to rule out such 'pathological' parameterizations, we impose a restriction on the distribution called *stability* . . . This restriction conveys the assumption that all the independencies imbedded in [the probability distribution] P are stable; that is, they are implied by the structure of the model D and hence remain invariant to any changes in the parameters [of D]. In our example only the correct structure (namely, $A \rightarrow C \leftarrow B$) will retain its independence pattern in the face of changing parameterizations – say, when one of the coins becomes slightly biased.[21]

We can see here two points of view that Pearl takes that make stability seem plausible to him. First, Pearl thinks causal systems should be decidable. It is clearly a criticism of the systems described that 'there is no way to decide among the three'. This attitude is revealed in discussions of other topics as well. For instance, as we shall see below, I reject the causal Markov condition. Pearl objects that by so doing I make questions about the causal structure and about the truth of certain counterfactuals unanswerable.[22]

Unanswerable given what information? Immediately after the section defining 'stability' Pearl tells us, 'With the added assumption of stability, every distribution has a unique minimal causal structure . . . as long as there are no hidden variables.'[23] Clearly he intends that the questions he is concerned about should be answerable given an order for the full set of causally relevant variables and the probability distribution over them. But so far as I can see, once we have given up the idea that there is something wrong with the notion of cause so that it has to be reduced away, there is no good reason to suppose that probabilities should be able to answer all questions about causality for us. (Nor am I sure that Pearl insists they should; for it is unclear whether he thinks all causal systems are stable, or takes the more modest line that his methods are capable of providing answers to all his questions only for systems that are stable.)

---

[21] Pearl (2000), p. 48.
[22] University of California at San Diego Philosophy of Economics Seminar, May 1999.
[23] Pearl (2000), p. 49.

The other point of view that matters for Pearl's claims about stability is the point of view of the engineer – which he is. It is apparent from the passage that Pearl thinks of causal structures as in some sense coming first: they get fixed, but then the parameter values can vary. But of course a causal system comes with both its structure and it parameters – you cannot have one without the other.

I think the way to put the response that makes sense of the idea of "structure first" is in terms of the kinds of operation we typically perform on the kinds of engineered devices Pearl generally has in mind. Think of a toaster. Its parts and their arrangement are fixed. We may bend the position of the trip plate a little, or of the expanding metal strip which it will meet, in order to keep the brownness of the toast calibrated with the settings on the brownness control. The values of the parameters do not matter so long as the basic causal structure does not break down; indeed the values are just the kind of thing we expect to drift over time. But we would have a legitimate cause of complaint if the same were true of the structure within the first year we owned the toaster.

That is fine for a toaster. But for other situations the parameters may matter equally with the structure, or more so. If birth-control pills do cause thrombosis we may work very hard to weaken the strength with which they do so at least to the point where people who take the pills are no worse off than those who do not. Indeed we may take this as an important aim – we are more obliged to get the effects to cancel out than we are to continue to spend money and research time to reduce the risk of thrombosis among pill takers below that of non-pill-takers. Getting the cancellation that stability/faithfulness prohibits is important to us.

This brings us to what seems to me a real oddity in the whole idea of stability/faithfulness. Probabilities and causal structures constrain each other. If the probability is fixed, then nature – and we – are not allowed to build certain kinds of causal structure. For instance, if we have the three binary variables, A, B and C, as in Pearl's example, with a probability in which they are pairwise independent and have to create a causal arrangement (or lack of!) among the three, we are prevented from building just the one Pearl describes. Or to think of it with causal structure first, as Pearl generally seems to, if C is to take value 1 or 0 depending on whether the outcome of the flip of two coins is the same or not, we are prevented from using fair coins and must introduce at least a little bias.

I come finally to the question of whether we should in fact expect to see a lot of causes that are not causes prima facie. A good many of the systems to which we think of applying the methods advocated by Bayes-nets theorists are constructed systems, either highly designed, like a toaster or an army admissions test, or a mix of intentional design, historical influence and unintended consequences, as in various socio-economic examples. In these cases cancellation of the effects of a given cause, either by encouraging the action of other factors or by encouraging the contrary operation of the cause itself, can be an important aim, particularly where the effect is deleterious. It will often be a lot easier to design for, or encourage the emergence of, cancellation than it is to eliminate the cause of the unwanted effect, or less costly or more overall beneficial (as in my discussion of birth-control pills). There is no good reason to assume that our aims are almost always frustrated.[24]

This is a view that Kevin Hoover also stresses in his work on causality in macroeconomics. He considers a macroeconomic example in which 'agency can result in constraints appearing in the data that [violate faithfulness]'.[25] He concludes:

Spirtes *et al.* . . . . acknowledge the possibility that particular parameter values might result in violations of faithfulness, but they dismiss their importance as having 'measure zero'. But this will not do for macroecomics. It fails to account for the fact that in macroeconomic and other control contexts, the policymaker aims to set parameter values in just such a way as to make this supposedly measure-zero situation occur. To the degree that policy is successful, such situations are common, not infinitely rare.[26]

Perhaps, however, the issue will be made: can we ever really expect exact cancellation? After all, to get an arrow in a Bayes-net causal graph, any degree of dependence between cause and effect will do. After we have the arrow in, we need not be misled by the smallness of the dependence to think the influence is small. For we can then insist on measuring degree of efficacy by the formula above that I and other causal decision theorists proposed and that Pearl endorses for $P(C\square \rightarrow E)$.

One reason we may think exact cancellations are rare is that actually getting any really precise value we aim for is rare. In a recent discussion of instrumentalism, Elliott Sober[27] talks about a comparison of the heights of corn plants in two populations. One thing we know, he claims, is that they are not really equal. Still, that is the working hypothesis. I take it one of the reasons he thinks we know this is that 'exactly equal' is very precise; and any very precise prediction is very likely to be wrong in an imprecise discipline.

This raises some very difficult issues about modelling and reality, especially for probabilities. We design a device to set the difference between two quantities at zero; tests for quality control show that, within bounds of experimental error, we succeeded; and we model the difference as zero. Should we think it 'really' is zero? It is not certain that the question makes sense, even when we are thinking of, say, a difference between the length of two strips in a single designated device. It becomes particularly problematic when we are thinking

---

[24] For further discussion see Cartwright (1999), ch. 2, and Cartwright (2000).
[25] Hoover (2001), pp. 7–33.   [26] Hoover (2001), pp. 7–35.   [27] Sober (1999).

about a difference of two probabilities in a population. Is the increase in probability of thrombosis on taking birth-control pills exactly offset by the decrease via pregnancy prevention in British women between the ages of 20 and 35 in the period from 1980 to 1990? All the conventional issues about what we intend by talking about the true probability become especially acute here.

Some, I think, we can sidestep, particularly when we are thinking about the application of the Bayes-nets approach to causality as opposed to the philosophical issue about substitutability raised by Spohn. For we are going to be using these methods in doing real social, medical and engineering science, using real data.[28] And here it is not unusual for our best estimates from the data to render two quantities probabilistically independent where estimates of appropriate partial conditional probabilities – as well perhaps as our background knowledge or even other kinds of tests we have conducted for the relevant causal connections – suggest the result is due to cancellation. In this case we either have to insist the probabilities are not those our best estimates indicate or forsake the commitment to faithfulness.

Before leaving this section I should repeat an old point, for completeness. Sometimes it is argued that Bayes-nets methods should supplement what we know. So if we do have independent evidence of cancellation, we should use it and not insist on faithfulness. But where we do not have such information we should assume faithfulness. As I indicated earlier, this strategy is ill founded; indeed, I think irresponsible. Where we don't know, we don't know. When we have to proceed with little information we should make the best evaluation we can for the case at hand – and hedge our bets heavily; we should not proceed with false confidence having plumped either for or against some specific hypothesis – like faithfulness – for how the given system works when we really have no idea.

6.4   From probabilistic dependence to causality

If we have a hypothesis that C causes E, we can use what we have just reviewed to test it, via the hypothetico-deductive method. But that is a method that we know to be more accurate at rejecting hypotheses than confirming them. Bayes-nets methods promise more: they will bootstrap from facts about dependencies and independencies to causal hypotheses – and, claim the advocates, never get it wrong.

---

[28] For an example of an attempt to use the Spirtes et al. methods on real economic data in economics, see Swanson and Granger (1997). Their struggles there are particularly relevant to my point in this paragraph. Which of the low partial correlations observed in their data should be taken to indicate that the 'true' partial correlation is zero? They consider various alternative choices among the lowest observed partial correlations and show that different choices give rise to different causal structures.

Again, as Spohn argues, if there really is this tight connection, there ought to be an argument for why it obtains. And there is. Again, we can see from looking at the argument why the inference from dependencies and independencies sometimes works, and why it will not work all the time. As with the other direction of inference, there is an argument for the connection and the argument itself makes clear that the connection is not tight.

What kinds of circumstances can be responsible for a probabilistic dependence between A and B? Lots of things. The fact that A causes B is among them: Causes produce their effects; they make them happen. So, in the right kind of population we can expect that there will be a higher frequency of the effect (E) when the cause (C) is present than when it is absent; and conversely for preventatives. With caveats.

What else? Here are a number of things, all discussed in the literature. (1) A and B may have a common cause or a common preventative or correlated causes or correlated preventatives, where either the causes are deterministic or the action of producing B occurs independently of the action producing A. (2) A and B may cooperate to produce an effect. In populations where the effect is either heavily present or heavily absent, A and B may be dependent on each other. (3) When two populations governed by different systems of causal laws or exhibiting different probability distributions are mixed together, the resulting population may not satisfy the causal Markov condition even though each of the subpopulations do. (This is analogous to Simpson's paradox reversals.) (4) A and B may be quantities with the same kind of temporal evolution, both monotonically increasing, say. Then the value of A at t will be probabilistically dependent on the value of B at t. (5) A and B may be produced as product and by-product from a probabilistic cause.

Let us look at each in turn and at what the defenders of Bayes nets have to say about them. I begin with (1) which is the case that advocates of Bayes-nets methods acknowledge and try to deal with squarely – assuming the underlying system is deterministic.

6.4.1 Why factors may be dependent

(1) Common causes (where nature is deterministic) Following Judea Pearl,[29] let us call the total effect of all those causes of X that are omitted from the variable set under consideration, V, and which combine with the direct causes in V of X to form a set of causes sufficient to fix the value of X, a random disturbance factor for X.[30] Bayes-nets methods are applied only to special sets of variables: sets V such that for each X in V, the random disturbance factor for

---

[29] Pearl (2000), p. 44.
[30] These are often designated u_x when variables in V are designated $x,y, \ldots$.

X is probabilistically independent of that for every other variable in V. In such a variable set we can prove that the causal Markov condition will be satisfied.[31]

The causal Markov condition, along with the assumption that all causes are prima facie causes, lies at the heart of the Bayes-nets methods. It tells us that a variable will be probabilistically independent of every other variable except its own effects once all of its direct causes have been conditioned on. So we eliminate cases where a dependence between A and B is due to reason (1) by requiring that the dependence persist once we have conditioned on the parents of A.

Everyone acknowledges that some constraint like this is necessary. You cannot get from dependence to causation; you at least have to first hold fixed the causal parents or something equivalent, then look for dependence. So in the remaining sections when I talk about the route from dependence to causation, I mean dependence conditional on a set of causal parents. To claim that this is enough to ensure a causal connection is to maintain the causal Markov condition.

My description of the restriction on the variable set V is rather long winded. The first reason is to avoid a small problem of characterization. What I have called 'random disturbance factors' are sometimes called 'exogenous' factors. There are various concepts of exogeneity. This usage obviously refers to the one in which exogenous factors are not caused by any variables in the system. Pearl clearly assumes that is true of random disturbance factors. But the proof requires more, for it is possible for all exogenous causes of one variable to be independent of those for another without the disturbance terms themselves being independent. That is because it is possible for a function of X and Y to be dependent on Z even if the three factors are pairwise independent. So it is not enough that the exogenous causes for a variable be independent of those for other variables: the proof needs their net effects to be independent.

The second reason is that some of the other terminology used in the discussion is unclear. Often we are told, as by Spirtes, Glymour and Scheines,[32] that the methods will be applied only to sets that are causally sufficient, adding the bold assumption that as a matter of empirical fact this will ensure the necessary independencies among the disturbance factors. But what is causal sufficiency? Spirtes et al. tell us, 'We require for causal sufficiency of V for a population that if X is not in V and is a common cause of two or more variables in V, that the joint probability of all variables in V be the same on each value of X that occurs in the population.'[33]

Let us assume that C is a common cause of A and B if C is a cause of A and a cause of B. The problem then is that this definition is too demanding.

[31] Cf. Verma and Pearl (1991) or Pearl (2000), p. 30.
[32] Cf. Spirtes et al. (1993), p. 54.   [33] Spirtes et al. (1993), p. 45.

Every cause of a common cause is itself a common cause. These could go back in time ad infinitum. And for any system for which there is a temporally continuous process connecting a cause with an effect at the type level, between each common cause and an earlier one there will be infinitely more. If we apply the methods only to variable sets that get them all in, we will not apply them at all. What we want to get in are all the last ones – the ones as close to both effects as possible.[34] But it will take some effort to formulate that properly. Spirtes et al. are particularly hampered here because they restrict their definitions to facts about causally correct representations rather than talking about causal relations in the world.

Spirtes, et al. avoid this problem by offering a different characterization. They define, 'We say that a variable X is a common cause of variables Y and Z if and only if X is a direct cause of Y relative to $\{X,Y,Z\}$ and a direct cause of Z relative to $\{X,Y,Z\}$'.[35] And for direct cause:

C is a direct cause of A relative to V just in case C is a member of some set **C** included in V/$\{A\}$ such that (i) the events in **C** are causes of A, (ii) the events in **C**, were they to occur, would cause A no matter whether the events in V/($\{A\} \cup$ **C**), were or were not to occur, and (iii) no proper subset of **C** satisfies (i) and (ii).[36]

The variables in V are seldom sufficient to fix the value of an effect. So how can fixing whether the events in **C** occur or not ensure that A occurs? So let us add into the set **C** the random disturbance factor for A. But in fact, it looks as if we have to assume as fixed all exogenous causes, or at least the 'last' one if that makes sense, since it will not help to fix one but allow temporally subsequent ones to vary. We also need to add that quantities occurring between **C** and A in nature's objective graph must be assumed to take on the values dictated by C. And so forth.

I do not know how to formulate all this correctly. But it needs to be done if the notion of causal sufficiency is to be used. Quite reasonably the advertisements for Bayes-nets methods make much of the fact that the subject is formal and precise: we can prove theorems about manipulation, about efficient conditioning sets for measuring the size of a causal effect, about the certainty of the results of the algorithms when applied to systems satisfying specific conditions, etc. But this is all pseudo-rationalism if we do not provide coherent characterizations of the concepts we are using.

The trouble with the characterization of 'causal sufficiency' arises from the fact that for Spirtes, Glymour and Scheines the notion of direct cause is relative to the choice of a particular variable set. Spohn's talk of the set of 'all the

---

[34] Note that where $A \ c \to B$ and $B \ c \to C$ and $B \ c \to D$, if C and D are independent conditioning on B, they need not be independent conditioning on A if $P(B/A) \neq 1$. So B must be included if the causal Markov condition is to be satisfied.

[35] Spirtes et al. (1993), p. 44.   [36] Spirtes et al. (1993), p. 43.

variables needed for a complete description of empirical reality',[37] temporally ordered, avoids this; Pearl, too, because he supposes that the underlying system is a set of deterministic causal laws on a finite set of causally ordered variables. No one to my knowledge has a good account of causal sufficiency for dense sets of effects, for instance, for the kinds of system studied by time-series analysis. As I remarked, Spirtes et al. talk only of correct causal representations.[38] That not only allows them to appear to avoid metaphysics, as Spohn and Pearl clearly do not, but also leaves an opening for supposing that the underlying metaphysics is continuous. But the advantages are illusory if we cannot produce adequate definitions.

Before proceeding to look at the list of factors that undermine the causal Markov condition, I should comment on one recent defence of it. Daniel Hausman and James Woodward[39] offer a proof of the condition alternative to the proof by Pearl and Verma. Central to their discussion is a concept they call modularity: each separate effect under study should be manipulable without disturbing any other. They claim that, given certain other conditions (such as the existence for each effect of a cause not in the variable set under consideration), the causal Markov condition is 'the flip side of' modularity. This would be a good defence if it were true. For we need not agree with Hausman and Woodward that all causal systems must be modular; we could nevertheless (supposing their other conditions are met) assume the causal Markov condition wherever we assume modularity.

The trouble is that the proof does not bear the interpretation they put on it. For given their other conditions, both modularity and the causal Markov condition follow separately. One is not the flip side of the other; both are the result of the conditions they suppose at the start. And these conditions are at any rate strong enough to call the Pearl and Verma proof of the causal Markov condition into play.[40]

6.4.2   Other reasons why factors may be dependent

(2) Two causes cooperating to produce one effect   When two causes cooperate to produce one effect, they will be mutually dependent in a population homogeneous with respect to that effect. These kinds of cases are common in practice. Data are hard to come by. We collect it for one reason but need to utilize it for many others. Imagine for example that we have data on patients

---

[37] Spohn (2001), p. 167.

[38] This is my account. They do not say what they do. For instance, in their section 'Axioms', they provide no axioms but only definitions (of the causal minimality condition, the faithfulness condition and the causal Markov condition). But I take it their claims are: every correct causal graph over a causally sufficient set of variables satisfies these conditions.

[39] Hausman and Woodward (1999).   [40] For a full discussion see chs. 8 and 9.

from a given hospital, where one disease, D, is especially prevalent. But we are interested in another condition, B. Unbeknown to us B cooperates with A in the production of D, so A and B are dependent in our population (even once we have conditioned on all the parents of A in a causally sufficient variable set). We erroneously infer that A causes B in this kind of population.

The problem here is not with the sample size. We can imagine that the sample is large and the frequencies are indicative of the 'true' probabilities for the population involved. The problem for the causal Markov condition is with the choice of population. We all know that to study the relation between A and B we should not use populations like this. But how should we – properly – characterize populations 'like this'?

(3) Mixing   Even if we assume the causal Markov condition for populations where the probabilities of the effect are fixed by the causal history, for mixed populations cooperating causes can still be correlated if the proportion of the effect is determined by some external factor rather than the causal history. Spirtes et al. tell us that there are no cases of mixing for causally sufficient variable sets: '*When a cause of membership in a sub-population is rightly regarded as a common cause of the variables in V, the Causal Markov Condition is not violated in a mixed population.*'[41] I think this is a bad idea: the 'variables' that are the 'cause of membership in a subpopulation' will often look nothing like variables – they do not vary in any reasonable way and there is no reason to think there is a probability distribution over them; and even if we did count them as variables, it looks as if we would have to count them as common causes of every variable in V to ensure restoration of the causal Markov condition.[42]

(4) Many quantities change in the same direction in time   There will thus be a probabilistic dependence between them. Social scientists solve this problem by detrending before they look for dependencies. Spirtes et al. maintain that there is no problem to solve. They use their previous solution to the problem of mixing plus a bold claim:

if we consider a series in which variable A increases with time, then A and B will be correlated in the population formed from all the units-at-times, even though A and B have no causal connection. Any such combined population is obviously a mixture of populations given by the time values.[43]

Like others[44] I find this claim ungrounded. Moreover it seems to me to be in tension with their commitment to determinism – which is important to them

---

[41] Spirtes et al. (1993), p. 60 (italics in original).
[42] For details see Cartwright (1999), ch. 5 and Cartwright (2000).
[43] Spirtes et al. (1993), p. 63.
[44] Cf. Berkovitz (2000) and Hoover (2001), ch. 7, and Sober (1988), pp. 161–2 and Sober (2000).

since in deterministic systems the causal Markov condition is bound to be true if only we add enough into the set of parents. Their idea I take it is that there will be different probability distributions across the causes operating at each time slice, hence mixing. But consider deterministic models in physics. These I take it are important for Spirtes, Glymour and Scheines because these are what make many people sympathetic to their claim that all macroscopic processes are deterministic. Any two systems moving inertially will have their positions correlated, and they have exactly the same causes operating at each instant with the same probability distribution over them, namely – none.

(5) *Products and by-products*   Products and by-products are mutually dependent, and when causes act purely probabilistically, no amount of conditioning on parents will eliminate the dependence. Perhaps then there is not any purely probabilistic causation – that would save the causal Markov condition.

But that is a big metaphysical gamble, especially in the face of the fact that for the kind of variables for which Bayes-nets methods are being sold, we seldom are able to formulate even a reasonable probabilistic model, let alone a deterministic one. We can of course stick to the metaphysical insistence that everything must be deterministic. I think this claim is unwarranted, but I will say no more about the matter here since I have written much about it elsewhere.[45]

6.5   Analysis

Dependence could be due to causation. But there are lots of other reasons for it as well. Bayes-nets methods stress one – the operation of common causes – and tell us how to deal with it when the underlying system is deterministic. The other four reasons standard in the philosophical literature are badly handled or made light of. And what about other reasons? Have we listed them all?

The reasons I listed are prevalent not only in the philosophical literature. They are standard fare in courses on social science methodology, along with lots of other cautions about the use of probabilities to infer causes in even experimental or quasi-experimental contexts. And they are not handled so badly there. In part the failures in the philosophical discussion arise from the requirement that the connection, whatever it is, be tight. We look for a claim of the form: A causes B iff A and B are probabilistically dependent in populations satisfying X. Then X is hard to formulate in the kind of vocabulary we need for formal proofs and precise characterizations.

But why should we think the connection is tight? As Spohn says, if it is tight there ought to be a reason. There is, as I have argued, a reason for the connection between probabilistic dependence and causality, but the very reason shows that

---

[45] See Cartwright (1997), Cartwright (1999), ch. 5 and Cartwright (2000).

the connection is not tight. Causes *can* increase the probability of their effects; but they need not. And for the other way around: an increase in probability *can* be due to a causal connection; but lots of other things can be responsible as well.

I think we are still suffering under the presumptions of the old Hume programme. First, we do not like modalities, especially strange ones. My breakfast cereal box says: 'Shredded Wheat can help maintain the Health of your Heart'. In the same sense, causes can increase the probability of their effects. Distressed at this odd modality,[46] we try to render this as a claim that causes *will* increase the probability of their effects, given X; then we struggle to formulate X. Second, we cannot get loose of the idea that causes need associations to make them legitimate. So we want some 'if and only if' with probabilities on the right, even if we grudgingly have to use some causal concepts to get the right-hand side filled in properly. I think we are looking at the issue entirely the wrong way. The connection between causes and probabilities is not like that. It is, rather, like the connection between a disease and one of its symptoms. The disease can cause the symptom, but it need not; and the same symptom can result from a great many different diseases.

This is why the philosopher's strategy is bad. We believe there must be some 'if and only if' and so are inclined to make light of cases that do not fit. The advice from my course on methods in the social sciences is better:

If you see a probabilistic dependence and are inclined to infer a causal connection from it, think hard about all the other possible reasons that that dependence might occur and eliminate them one by one. And when you are all done, remember – your conclusion is no more certain than your confidence that you really have eliminated all the possible alternatives.

---

[46] I offer a treatment – though not yet really satisfactory – of these kinds of modality in Cartwright (1999).

7   Modularity: it can – and generally does – fail

7.1   Introduction

This chapter pursues themes developed in my recent book, *The Dappled World: a Study of the Boundaries of Science.*[1] The book is a Scotist book – in accord with the viewpoint of Duns Scotus. It extols the particular over the universal, the diverse over the homogeneous and the local over the global. Its central thesis is that in the world that science studies, differences turn out to matter. Correlatively, universal methods and universal theories should be viewed with suspicion. We should look very carefully at their empirical justification before we adopt them.

The topic in the volume where this chapter was first published is causality; I shall defend a particularist view of our subject. Causal systems differ. What is characteristic of one is not characteristic of all and the methods that work for finding out about one need not work for finding out about another. I shall argue this here for one specific characteristic: modularity. Very roughly, a system of causal laws is modular in the sense I shall discuss when each effect in the system has one cause all its own, a cause that causes it but does not cause any other effect in the system. On the face of it this may seem a very special, probably rare, situation. But a number of authors currently writing on causality suppose just the opposite. Modularity, they say, is a universal characteristic of causal systems. I shall argue that they are mistaken.

7.2   What is modularity?

Behind the idea that each effect in a causal system[2] should have a cause of its own is another idea, the idea that each effect in the system must be able to take any value in its range consistent with all other effects in the system taking any values in their ranges. There are two standard ways in which people seem

---

[1] Cartwright (1999).
[2] I shall use 'causal system' to refer to a set of causal laws and 'causal structure' to refer to a set of hypotheses about causal laws.

to think this can happen; it will be apparent that different senses of 'able' are involved.

In the first place, a second collection of causal systems very similar to the first may be possible in which all the laws are exactly the same except for the laws for the particular effect in question. In the new systems these laws are replaced by new laws that dictate that the effect take some specific value, where the systems in the collection cover all the values in the range of that effect.

This interpretation clearly requires that we be able to make sense of the claim that an alternative set of laws is possible. For my own part I have no trouble with this concept: in *The Dappled World* I argue that laws are not fundamental but instead arise as the result of the successful operation of a stable arrangement of features with stable capacities. Nevertheless, I do not see any grounds for the assumption that the right kind of alternative arrangements must be possible to give rise to just the right sets of laws to make modularity true. At any rate this way of securing modularity is not my topic in this chapter.

The second way in which modularity might obtain is when each effect in the system has a cause all of its own that can contribute to whatever its other causes are doing to make the effect take any value in its range. This is the one I will discuss here. I will also in this chapter restrict my attention to systems of causal laws that are both linear and deterministic. In this case the commitment to modularity of the second kind becomes a commitment to what I call 'epistemic convenience'.

An epistemically convenient linear deterministic system is a system of causal laws of the following form[3]

$$x_1 \, c^= u_1$$
$$x_2 \, c^= a_{21}x_1 + u_2$$
$$\vdots$$
$$x_n \, c^= \Sigma a_{nj}x_j + u_n$$

plus a probability measure $P(u_1, \ldots, u_n)$, where
(i) there are no cross restraints among the u's;[4] and the u's are probabilistically independent of each other;
(ii) for all j, Prob $(u_j = 0) \neq 1$.
The symbol $c^=$ shows that the law is a causal law. It implies both that the relation obtained by replacing $c^=$ with '=' holds and that all the quantities referred to on the right-hand side are causes of the one on the left-hand side. The u's will

---

[3] Somewhat more accurately, I should say 'a system of laws generated by laws of the following form', for I take it that causal laws are transitive. For a more precise formulation, see ch. 10 in this book.

[4] For a discussion, see Cartwright (1989).

be hence forward referred to as 'exogenous': they cause quantities in the set of factors under study (represented by $x_1, \ldots x_n$) but are not caused by them.

These systems are epistemically convenient because they make it easy to employ randomized treatment/control experiments to settle questions about causality using the method of concomitant variation. I will explain in more detail below, but the basic idea can be seen by considering the most straightforward version of the method of concomitant variation: to test if x_j causes x_e and with what strength, use u_j to vary x_j while holding fixed all the other u's, and look to see how x_e varies in train. Conditions (i) and (ii) guarantee that this can be done.

A number of authors from different fields maintain that modularity[5] is a universal characteristic of causal systems. This includes economic methodologist Kevin Hoover,[6] possibly Herbert Simon,[7] economists T. F. Cooley and Stephan LeRoy,[8] Judea Pearl in his new study of counterfactuals,[9] James Woodward,[10] Daniel Hausman[11] and Daniel Hausman and James Woodward[12] jointly in a paper on the causal Markov condition. I aim to show that, contrary to their claims, we can have causality without modularity. I focus on the second kind of modularity here in part because that is the kind I have found most explicitly defended. Hence I shall be arguing that not all causal systems are epistemically convenient.[13]

7.3   The method of concomitant variation

We say that the method of concomitant variation is a good way to test a causal claim. But can we show it? For an epistemically convenient system we can, given certain natural assumptions about causal systems. That is one of the best things about an epistemically convenient system – we can use the method of concomitant variation to find out about it.

---

[5] Or some closely related doctrine. Much of what I say can be reformulated to bear on various different versions of a modularity-like condition.

[6] See his defence of the invariance of the conditional probability of the effect on the cause in Hoover (2001). In this discussion Hoover seems to suppose that there always is some way for a single cause to vary and to do so without any change in the overall set of laws. At other places, I think, he does not assume this. But he does speak with approval of Herbert Simon's way of characterizing causal order, and Simon's characterization requires the possibility of separate variation of each factor.

[7] Simon (1957).   [8] Cooley and LeRoy (1985).   [9] Pearl (2000).

[10] Woodward (1997).   [11] Hausman (1998).   [12] Hausman and Woodward (1999).

[13] The authors mentioned here all have slightly different views, formulated and defended differently and with different caveats. I apologize for lumping them all together. Clearly not all the remarks I make are relevant to every view. In fact I will focus on a very specific form of the claim for universal modularity. Nevertheless, most of what I say can be translated to apply to other forms of the claim.

I shall not give the proof here, but rather describe some results we can show.

Here are the assumptions I shall make about linear deterministic systems of causal laws:

A_1 *Functional dependence*: any causal equation presents a true functional relation.

A_2 *Anti-symmetry and irreflexivity*: if q causes r, r does not cause q.

A_3 *Uniqueness of coefficients*: no effect has more than one expansion in the same set of causes.

A_4 *Numerical transitivity*: causally correct equations remain causally correct if we substitute for any effect any function in its causes that is among nature's causal laws.

A_5 *Consistency*: any two causally correct equations for the same effect can be brought into the same form by substituting for effects in them functions of their causes given in nature's causal laws.

A_6 *Generalized Reichenbach principle*: no quantities are functionally related unless the relation follows from nature's causal laws.

The result I shall describe says very roughly that when the underlying linear deterministic system is an epistemically convenient one, then a causal hypothesis is correct iff the method of concomitant variation says it is so. In order to express this more precisely, we shall have to know what the form of the causal hypotheses in question are, what it is for hypotheses of this form to be causally correct and what it is to pass the test for concomitant variation.

The usual hypotheses on offer when we suppose the underlying causal system to be an epistemically convenient linear deterministic system[14] are in the form of regression equations:

$$\mathbf{R}: x_k \stackrel{c}{=} \Sigma a_{kj}x_j + \Psi_k, \text{ for } \Psi \perp x_j \text{ for all } j$$

where $x \perp y$ means that $\langle xy \rangle = \langle x \rangle \langle y \rangle$.

What exactly does this hypothesis claim? I take it that the usual understanding is this: every quantity represented by a named variable (an x) on the right-hand side is a genuine cause of the quantity represented on the left-hand side, and the coefficients are 'right'. The random variable Ψ represents a sum of not-yet-known causes that turn $\mathbf{R}$ into a direct representation of one of the laws of the system. So I propose to define correctness thus. An equation of the form $\mathbf{R}$: supposing the variable x_j represents the quantity q_j, $x_k \stackrel{c}{=} \Sigma a_{kj}x_j + \Psi$ $(1 \leq j \leq m)$, for $\Psi_k \perp x_j$, is correct iff there exist $\{b_j\}$ (possibly $b_j = 0$), $\{q_j'\}$ such that $q_k \stackrel{c}{=} \sum a_{kj}q_j + b_jq_j' + u_k$, where q_j does not cause q_j'. (This last restriction ensures that all the omitted factors are causally antecedent to or 'simultaneous' with those mentioned in the regression formula.)

---

[14] Or when we are prepared to model the system as linear and deterministic for some reason or another.

Now let us consider concomitant variation. In an epistemically convenient linear deterministic system, the values of the x's are fixed by the u's, and the u's can vary independently of each other. The core idea of the method is to take the concomitant variation between x_c and x_e when u_c varies while all the other u's are fixed as a measure of the coefficient of x_c in nature's equation for x_e.

To state the relevant theorems we shall need some notation. Let

$$\Delta j(\alpha)x_n =_{df} x_n(u_1 = U_1, \ldots, u_{j-1} = U_{j-1}, u_j \\ = U_j + \alpha, u_{j+1} = U_{j+1}, \ldots, u_m \\ = U_m) - x_n(u_1 = U_1, \ldots, u_{j-1} \\ = U_{j-1}, u_j = U_j, u_{j+1} = U_{j+1}, \ldots, u_m = U_m)$$

Then we can prove the following.[15]

Theorem 1 A (true) regression equation $x_k \,^c = \Sigma_{j=1}^{k-1} a_{kj}x_j + \Psi_k$, is causally correct iff for all values of α and J, $1 \leq J \leq k$, $\Delta_J(\alpha)x_k = \Sigma a_{kj} \Delta_J(\alpha)x_j$; i.e. iff the equation predicts rightly the differences in x_k generated from variations in any right-hand-side variable.

Notice, however, that this theorem is not very helpful to us in making causal inferences because it will be hard to tell whether an equation has indeed predicted the differences rightly. That is because we will not know what $\Delta_J(\alpha)x_j$ should be unless we know how variations in u_J affect x_j and to know that we will have to know the causal relations between x_J and x_j. So in order to judge whether each of the x_j affects x_k in the way hypothesized, we will have already to know how they affect each other. If we happen to know that none of them affects the others at all, we will be in a better situation, since the following can be trivially derived from the previous theorem.

Theorem 2 A (true) regression equation for x_k in which no right-hand-side variable causes any other is causally correct iff for all α and J, $\Delta_J(\alpha)x_k = a_{kJ} \Delta_J(\alpha)u_J$.

We can also do somewhat better if we have a complete set of hypotheses about the right-hand-side variables. To explain this, let me define a complete causal structure that represents an epistemically convenient linear deterministic system with probability measure, as a triple, $\langle X = \{x_1, \ldots, x_n\}, \mu, \text{CLH}\rangle$, where μ is

[15] For a proof of the three theorems see ch. 10 in this book. The formalization and proofs are inspired by the work of James Woodward on invariance, which argues more informally for a more loosely stated claim. I am aiming to make these claims more precise in these theorems.

a probability measure over X and where the causal law hypotheses, CLH, have the following form:

$$x_1 \, c^= \Psi_1$$
$$x_2 \, c^= a_{21}x_1 + \Psi_2$$
$$\vdots$$
$$x_n \, c^= \Sigma_{j=1}^{n-1} a_{nj} x_j + \Psi_n$$

with $\Psi_j \perp x_k$, for all $k < j$. In general $n < m$, where m is the number of effects in the causal system. Now I can formulate the following.

Theorem 3 If for all x_k in a complete causal structure, the $\Delta_J(\alpha)x_k$ that actually obtains equals $\Delta_J(\alpha)x_k$ as predicted by the causal structure for all α and J, $1 \leq J \leq n$, then all the hypotheses of the structure are correct.

I take it that it is the kind of fact recorded in these theorems that make epistemically convenient systems so desirable, so that we might wish – if we could have it – for all causal systems to be epistemically convenient. But is it sensible to think they are? In the next section I will give some obvious starting reasons for thinking the answer must be 'no'.

7.4   Three peculiarities of epistemic convenience

To notice how odd the requirement of epistemic convenience is, let us look first at some ordinary object whose operation would naturally be modelled at most points by a system of deterministic laws – for instance a well-made toaster like the one illustrated in David Macaulay's *How Things Work*. The expansion of the sensor due to the heat produces a contact between the trip plate and the sensor. This completes the circuit, allowing the solenoid to attract the catch, which releases the lever. The lever moves forward and pushes the toast rack open.

I would say that the movement of the lever causes the movement of the rack. It also causes a break in the circuit. Where then is the special cause that affects only the movement of the rack? Indeed, where is there space for it? The rack is bolted to the lever. The rack must move exactly as the lever dictates. So long as the toaster stays intact and operates as it is supposed to, the movement of the rack must be fixed by the movement of the lever to which it is bolted.

Perhaps, though, we should take the bolting of the lever to the rack as an additional cause of the movement of the rack? In my opinion we should not. To do so is to mix up causes that produce effects within the properly operating toaster with the facts responsible for the toaster operating in the way it does; that is, to confuse the causal laws at work with the reason those are the causal laws at work.[16] But even if we did add the bolting together at this point as a cause, I do not see how it could satisfy conditions (i) and (ii) (in section 7.2). It does after all happen as part of the execution of the overall design of the toaster, and hence it is highly correlated with all the other similar causes that we should add if we add this one, such as the locating of the trip plate and the locating of the sensor.

The second thing that is odd about the demand for modularity is where it locates the causal nexus. It is usual to suppose that the fact that C causes E depends on some relations between C and E.[17] Modularity makes it depend on the relation between the causes of C and C: C cannot cause anything unless it itself is brought about in a very special way.

Indeed, I think that Daniel Hausman embraces this view:

> people . . . believe that causes make their effects happen and not vice versa. This belief is an exaggerated metaphysical pun, which derives from the fact that people can make things happen by their causes. This belief presupposes the possibility of intervention and the claim that not all the causes of a given event are nomically connected to one another.[18]

This is a very strong view that should be contrasted with the weaker view, closer (on my reading) to that of Hume, that the concept of causation arises because 'people can make things happen by their causes', but that this condition does not constitute a truth condition for causation. The weaker view requires at most that sometimes a cause of a cause of a given effect vary independently of all the 'other' causes of that effect; it does not take epistemic convenience as universal.[19] In my opinion the weaker view only makes sense as an empirical or historical claim about how we do in fact form our concepts, and about that, we still do not have a reliable account. The stronger view just seems odd.

Thirdly, the doctrine seems to imply that it is impossible to build a bomb that cannot be defused. Nor can we make a deterministic device of this sort: the

---

[16] For a more complete discussion of this point, see the distinction between nomological machines, on the one hand, and the laws that such machines give rise to, on the other, in Cartwright (1999).

[17] Or perhaps, since C and E here pick out types and not particular events, 'between C-type events and E-type events'.

[18] Hausman (1998).

[19] It is also surprising that Hausman focuses on the supposition of (something like) epistemic convenience as a necessary condition, but does not stress the equally problematic matter of the possibility of choice. We all know the classic debate about free will and determinism: it looks as if people cannot make things happen by their causes unless the causes of the causes are not themselves determined by factors outside our will, and that in turn looks to preclude universal determinism. If that should follow, it would not trouble me, but many advocates of modularity also defend the causal Markov condition – which I attack – on the grounds of universal determinism. Moreover, the need for us to cause some of the causes at least some of the time seems equally necessary whether one takes the strong view that Hausman maintains or the weaker view – which does not require epistemic convenience as a truth condition – that I described as closer to Hume.

correct functioning of the mechanisms requires that they operate in a vacuum; so we seal the whole device in a vacuum in such a way that we cannot penetrate the cover to affect one cause in the chain without affecting all of them. Maybe we cannot build a device of this sort – but why not? It does not seem like the claim that we cannot build a perpetual motion machine. On the doctrine of universal epistemic convenience we either have to say that these devices are indeed impossible, or that what is going on from one step to the next inside the cover is not causation, no matter how much it looks like other known cases of causation or passes other tests for causation (such as the transfer of energy/momentum test or the demand that a cause increase the probability of its effects holding fixed a full set of other causes).

Given that the claim to epistemic convenience as a universal condition on causality has these odd features, what might motivate one to adopt it? Three motivations are ready to hand: we might be moved by operationalist intuitions, or by pragmatist intuitions or we might be very optimistic about how nicely the world is arranged for us. I will take up each in turn.

7.5   Motivations for epistemic convenience: 'excessive' operationalism

This is a hypothesis of Arthur Fine's:[20] advocates of modularity conflate the truth conditions for a causal claim with conditions which were they to obtain would make for a ready test. As we have seen, a central feature of deterministic systems that are epistemically convenient is that we can use the simplest version of the method of concomitant variation within them: to test 'x_c causes x_e', consider situations in which x_c varies without variation in any 'other' causes of x_e and look for variation in x_e. I think this is particularly plausible as a motivation for economists. Economists in general tend to be very loyal to empiricism, even to the point of adopting operationalism. For instance, they do not like to admit preferences as a psychological category but prefer to use only preferences that are revealed in actions.

In general, versions of operationalism that elevate a good test to a truth condition are in disfavour. Still we need not dispute the matter here, for, even were we disposed to this kind of operationalism in the special case at hand, it would not do the job. Simple concomitant variation is no better test than many others – including more complicated methods of concomitant variation. So operationalism will not lead us to limit causal concepts to systems that admit tests by simple concomitant variation at the cost of other kinds of system. In particular, the simple method does not demand any 'less'[21] background

---

[20] Conversation, May, 2000, Athens, Ohio.

[21] I put 'less' in scare quotes because I do not mean to get us involved in any formal notions of more and less information.

knowledge than tests using more complicated versions of concomitant variation, which can be performed on other kinds of deterministic systems, or knowledge of a different kind.

Let me illustrate. We will continue to look at linear deterministic systems and we will still assume that all exogenous factors are mutually unconstrained: there are no functional relations between them.[22] And we will still test for causal relations by the method of concomitant variation.

Imagine then that we wish to learn the overall strength, if any, of x_1's capacity to affect x_e, where we assume we know some cause u_1 that has a known effect (of, say, size b_1) on x_1 and whose variation we can observe. In the general case where we do not presuppose epistemic convenience, every candidate for u_1 may well affect x_e by other intermediaries, say $x_2, \ldots, x_n$, as well. Suppose the overall strength of its capacity to affect x_j is b_j and of x_j to affect x_e is c_j.

We aim to compare two different situations, which are identified by the values assigned to the u's: $S = \langle U_1, U_2, \ldots U_m \rangle$ and $S' = \langle U_1', U_2, \ldots, U_m \rangle$, where the u's constitute a complete set of mutually unconstrained exogenous factors that determine x_e. Then

$$x_e' - x_e = \Sigma_{j=1}^{m} b_j c_j (U_1' - U_1)$$

or, letting $(\Sigma_{j=2}^{m} b_j c_j)/b_1 = df A$

$$(x_e' - x_e)/b_1(U_1' - U_1) = c_1 + A$$

Now here is the argument we might be tempted to give in favour of epistemically convenient systems. If we have an epistemically convenient system, $A = 0$, so $c_1 = x_e' - x_e/b_1 (U_1' - U_1)$. Otherwise we need to know the value of A in order to calculate c_1, the strength of x_1's capacity to affect x_e. So we need less antecedent knowledge if our systems are epistemically convenient.

But clearly the last two sentences are mistaken: $A = 0$ is just as much a value of A as any other; to apply the method of concomitant variation, we need to know (or be willing to bet on) the value of A in any case. Sometimes there may be some factor u_1 for which it is fairly easy to know that its effect on x_e by routes other than x_1 is zero. This for example, is, in my opinion, the case with J. L. Mackie's famous hypothesis that the sounding of the end-of-workday hooters in Manchester brings the workers out on to the streets in London. Here we know various ways to make the hooters in Manchester sound of which we can be fairly confident that they could not get the workers in London out except via making the Manchester hooters hoot.[23]

---

[22] If they are constrained, re-express all of them as appropriate functions of a further set of mutually unconstrained factors. Notice that this has no implications one way or another about whether, for example, two endogenous factors share all the same exogenous causes. The point of this is to allow that the exogenous factors can vary independently of each other.

[23] For a discussion, see Cartwright (1989).

But equally, sometimes we may know for some exogenous factors that do affect x_e by routes other than x_1 what the overall strength of that effect is – if, for instance, we have data on variations in x_e given variations in u_1 when the route from u_1 to x_1 is blocked.

Let us review some of the prominent facts we would need to know for a brute-force application of the method of concomitant variation, as I have described it, in a linear deterministic system. To test 'x_1 causes x_e with strength c' we need to know:

1 of a factor u_1 that it is exogenous to the system under study, that it causes x_1 and with what strength it does so;

2 of a set of factors that they are exogenous, that they are mutually unconstrained, and that together, possibly including u_1, they are sufficient to fix x_e but not sufficient by themselves to fix u_1;

3 what would happen to x_e in two different situations for which the values of the exogenous factors described in (2) do not vary, except for the value of u_1, which does vary;

4 the overall strength of u_1's capacity to affect x_e by other routes than by causing x_1.

My point here is that we need to know (or find a way around knowing) all of this information whether or not the system is epistemically convenient.

Why then have I called these special kinds of system 'epistemically convenient' for use of the method of concomitant variation if we need to know (or find our way around knowing) 'the same amount' of information to use the method whether the system is epistemically convenient or not? Because when the system is epistemically convenient, it is a lot easier to use randomized treatment/control experiments. That is why I have called these systems 'epistemically convenient'; and it is one of the chief arguments James Woodward gives in favour of the claim that causal systems should be epistemically convenient:

A manipulationist approach to causation explains the role of experimentation in causal inference in a very simple and straightforward way: Experimentation is relevant to establishing causal claims because those claims consist in, or have immediate implications concerning, claims about what would happen to effects under appropriate manipulations of their putative causes. In other words, the connection between causation and experimentation is built into the very content of causal claims.[24]

Randomized treatment/control experiments provide us with a powerful tool to find our way around knowing large chunks of information we would otherwise need to know. For the point at issue in this chapter, we need to be clear about which features of the stock experimental structure help with which aspects of our ignorance.

[24] Woodward (2000).

Randomization allows us to finesse our lack of knowledge of the kinds of facts described in (2) above. When we are considering the effect of u_1 on x_e, we generally do not know a set of 'other' exogenous factors sufficient to fix x_e. But a successful randomization ensures that they will be equally distributed in both the treatment and the control groups. Hence there will be no background correlations between these other factors that might confound our results. Observing the outcome in the two groups allows us to find out (roughly)[25] the information we look for in (3): what happens under variation in u_1.

But notice that randomization and observation do these jobs whether or not the system is epistemically convenient. Epistemic convenience matters because we were trying to find out, not about the effects of u_1, but rather about the effects of x_1. In the case I described above, where economic convenience fails, u_1 has multiple capacities: it can affect x_e differently by different routes. We are interested only in its effect via x_1, which we shall use to calculate the effect of x_1 itself.[26] Randomization does not help with this problem. Just as in the brute-force application of concomitant variation, we need either to find a cause of x_1 which we know has no other way of affecting x_e, or we need to know the overall effect via other routes in order to subtract it away.

The placebo effect is a well-known example of this problem. Getting the experimental subjects to take the medicine not only causes them to have the medicine in their bodies. It can also affect recovery by producing various psychological effects – feeling cared for, optimism about recovery, etc.

This is a good example to reflect on with respect to the general question of how widespread epistemically convenient systems are. How do we canonically deal with the placebo effect? We give the patients in the control group some treatment that is outwardly as similar to the treatment under test as possible but that is known to have no effect on the outcome under study.

That is, we do not hunt for yet another way to get the medicine into the subjects, a way that does not affect recovery by any other route. Rather we accept that our methods of so doing may affect recovery in the way suggested (or by still other routes) and introduce another factor into the control group that we hope will just balance whatever these effects (if any) may be. Ironically then, the standard procedure in these medical experiments does not support the claim

---

[25] The experiment does not allow us to tell what happens for any two specific situations (i.e. any specific choice of values for the u_j) but only certain coarser facts. For instance, if u_1 is causally unanimous across all situations (i.e. it is either causally positive across all, or causally negative or causally neutral), for a two-valued outcome, x_e, it can be shown that the probability of x_e in the treatment group is respectively greater than, less than or equal to that in the control group iff u_1 is causally positive, negative or neutral with respect to x_e.

[26] It may be just worth reminding ourselves, so as not to confuse the two issues, that x_1 itself may have multiple capacities with respect to x_e. Simple randomized treatment/control experiments do not disentangle the various capacities of x_1 but rather teach us about the overall effect of x_1 on x_e.

that there is always a way to manipulate the cause we want to test without in any other way affecting the outcome. Epistemic convenience definitely makes randomized treatment/control experiments easier, but there are vast numbers of cases in which we do not rely on it to hold.

7.6   Motivations for epistemic convenience: 'excessive' pragmatism

This is a hypothesis raised by the students in my doctoral seminar on causality in economics at LSE: advocates of modularity elevate a plausible answer to the question 'Of what use to us is a concept of causation?' into a truth condition. This motivation is explicitly acknowledged by Daniel Hausman in defense of a similar condition to the one we are investigating in his book *Causal Asymmetries*:

> What do people need causal explanations or a notion of causation for? Why isn't it enough to know the lawlike relations among types? Because human beings are actors, not just spectators. Knowledge of laws will guide the expectation of spectators, but it does not tell actors what will result from their interventions. The possibility of abstract intervention is essential to causation . . .[27]

My remarks here are identical to those about operationalism. Whether or not we wish to adopt the pragmatic justification as a truth condition, it will not do the job of defending modularity as a truth condition. Consider the same example as above. The conditions for using variations in u_1 to produce variations in x_1 and thereby to obtain predictable variations in x_e are much the same as the conditions for testing via concomitant variation.

To know what we will bring about in x_e by manipulating u_1 it is not enough to know just the influence of u_1 on x_1 and of x_1 on x_e. We also need to know the overall influence of u_1 on x_e by all other routes. Knowing that the size of influence by other routes is zero is just a special case. Whatever its value, if we know what the value is we can couple this knowledge with our knowledge of the influence of u_1 via x_1 to make reliable predictions of the consequences of our actions. So we do not need modularity to make use of our causal knowledge.

There is, however, a venerable argument for a different conclusion lurking here. If we are to use our causal knowledge of the link from x_1 to bring about values we want for x_e, it seems that some cause or other of x_1 must not itself be deterministically fixed by factors independent of our wishes: there must be some causes of x_1 that we can genuinely manipulate. But again, whether or not this is a good argument, it does not bear on modularity. To make use of our knowledge of the causal link between x_1 and x_e we may need a cause of x_1

---

[27] Hausman (1998), pp. 96–7.

that we can manipulate; but that does not show that we need a cause we can manipulate without in any other way affecting x_e.

7.7   Motivations for epistemic convenience: 'excessive' optimism

This is my hypothesis of why many are eager to believe that epistemic convenience is at least widespread, if not universal. Life becomes easier in a number of ways if the systems we study are epistemically convenient. Statistical inference of the strengths of coefficients in linear equations can become easier in well-known ways. So too can causal inference, in ways I have discussed here. And, as we shall see, Judea Pearl can provide a very nice semantics for counterfactuals as well as for a number of other different causal notions. Wishful thinking can lead us to believe that all systems we encounter will meet the conditions that make life easier. But wishful thinking must be avoided here, or we will be led into the use of methods that cannot deliver what we rely on them for.

I think we can conclude from these considerations that these three motivations do not provide strong enough reason to accept universal epistemic convenience. What positive arguments then are on offer on its behalf?

7.8   For and against epistemic convenience

7.8.1   Hausman's claim

Daniel Hausman points out that the cause we focus on is not generally the complete cause. A complete cause will include both helping factors and the absence of disturbances. Even if effects share the causes we normally focus on (e.g. in the toaster as I described it, the breaking of the circuit and the moving of the rack are both caused by the motion of the lever), they will not share all of these other factors, Hausman maintains.

Disturbing factors This claim seems particularly plausible with respect to disturbing factors. Most of the effects we are modelling here are fairly well separated in time and space. So it seems reasonable to expect that some things that might disturb the one would not disturb the other. This seems promising for the thesis of, if not universal, at least widespread, epistemic convenience. But there is a trouble with disturbing factors: often what they do is to disrupt the relation between the causes and the effect altogether. To salvage epistemic convenience, they need instead to cooperate with the causes adding or subtracting any spare influence necessary to ensure that the effect can take all the values in its allowed range. So they do not seem to satisfy reliably the conditions for epistemic convenience.

Helping factors   Return to the toaster. The motion of the lever causes the motion of the rack. That of course depends on the fact that the lever is bolted solidly to the rack: if the lever were not bolted to the rack, the lever could not move the rack. Could we not then take the fact that the lever is bolted to the rack to be just what we need for the special cause of the motion of the rack, a cause that the motion of the rack has all to itself?

I think not, for a number reasons:

1 As I urged in section 7.4, the fact that the two are bolted together is not one of the causes within the system of causal laws but rather part of the identification of what that systems of laws applies to, and this identification matters. We do not, after all, seek to know what the causal law is that links the movement of levers in general with the movement of racks of the right shape to contain toast. Surely there is no such law. Rather we want to know the causal relation, if any, between the movement of the lever and the movement of the rack in a toaster of this particular design. Without a specific design under consideration, the question of the causal connection, or lack of it, between levers and racks is meaningless.

2 Let us, however, for the sake of argument, admit as a helping cause in the laws determining the motion of the rack the fact that the lever and rack are bolted together. My second worry about calling on helping factors like this to save epistemic convenience depends on the probability relations these factors must bear to each other. In section 7.4, I queried whether these factors would be probabilistically independent of each other. Here I want to ask a prior question. Where is the probability distribution over these factors supposed to come from and what does it mean?

We could consider as our reference class toasters meeting the specific set of design requirements under consideration. Then the probabilities for all of these 'helping factors' being just as they are could be defined and would be 1. Independence would be trivially obtained, but at the cost of the kind of variation we need in the values of the u's to guarantee, via our theorems, that concomitant variation will give the right verdicts about causality.

Alternatively the reference class could be the toasters produced in a given factory following the designated design. Presumably then there would be some faults some time in affixing the lever to the rack so that not all the u's would have probability 1. But will the faults be independent? If not this reference class, then what? It will not do to have a make-believe class, for how are we to answer the question: if the attachment of the lever to the rack were to vary, what would happen to the rack? We need some other information to tell us that – most usually, I would suppose, knowledge of the causal connections in the toaster! And if not that information exactly, I bet it would nevertheless be information sufficient to settle the causal issue directly, without detour through concomitant variations.

3 The third worry is about the range of variation. For the theorems to guarantee the reliability of the method of concomitant variation, we need u's that will take the cause under test through its full range relative to the full range of values for the other causes. Otherwise there could be blips – the causal equation we infer is not true across all the values but depends on the specific arrangement of values we consider. Will the factors we pick out have a reasonable range of variation? This remark applies equally well to disturbing factors.

4 Last I should like to point out two peculiarities in the way people often talk about the factors designated by the u's. Often they are supposed not only to represent the special causes peculiar to each separate effect but also all the 'unknown' factors we have not included in our model. But if they are unknown, we can hardly use them as handles for applying the method of concomitant variation. And if epistemically convenient systems are not going to be of epistemic convenience after all, why should we want them? I realize that the issue here is not supposed to be whether we want systems of this kind, but rather whether we have them. Still in cases like this where the answer is hard to make out, the strategy should be to ask what depends on the answer. That is the reasonable way to establish clear criteria for whether a proffered answer is acceptable or not.

The second peculiarity arises from talking of the u's as a 'switch' that turns the cause to different values. Often it is proposed that the switch is usually 'off' yet could be turned on to allow us to intervene. This raises worries about the independence requirements on the u's again. Why should that kind of factor have a probability distribution at all, let alone one that renders it independent of all the other switch variables?

Judea Pearl and modularity Judea Pearl supposes that modularity holds in the semantics he provides for singular counterfactuals. He claims that, without modularity, counterfactuals would be ambiguous.[28] So modularity must obtain wherever counterfactuals make sense. This will double as an argument for universal modularity if we think that counterfactuals make sense in every causal system.

Pearl assumes modularity of the first kind, where alternative causal systems of just the right kind are always possible, but I can explain something of how the semantics works using the epistemically convenient systems we have been studying here. We ask, for instance, in a situation where $x_j = X_j$ and $x_k = X_k$, 'Were $x_j = X_j + \Delta$, would $x_k = X_k + a_{kj}\Delta$?' The question may be thought ambiguous because we do not know what is to stay fixed as x_k varies. Not so if we adopt the analogue of Pearl's semantics for our epistemically convenient

---

[28] 14 March 2000, seminar presentation, Department of Philosophy, University of California at San Diego.

system. In that case u_j must vary in order to produce the required variation in x_j and all the other u's must stay the same.

The semantics Pearl offers is very nice, but I do not see how it functions as an argument that counterfactuals need modularity. The counterfactuals become unambiguous just because Pearl provides a semantics for them and because that semantics always provides a yes–no outcome. Any semantics that does this will equally make them unambiguous.

Perhaps we could argue on Pearl's behalf that his is the right semantics, and it is a semantics that is not available in systems that are not epistemically convenient. Against that we have all the standard arguments that counterfactuals are used in different ways, and Pearl's semantics – like others – only accounts for some of our uses. We should also point out that we do use, and seem to understand, counterfactuals in situations where it is in no way apparent that the causal laws at work are epistemically convenient.

I think one defence Pearl may have in mind concerns the connection between counterfactuals and causality. Consider a very simple case where one common cause, v, is totally responsible both for x_1 and x_2 and no u_1 is available to vary x_1 independently of v.[29]

It is easy to construct a semantics, similar indeed to the one Pearl does construct, that answers unambiguously what would happen to x_2 if x_1 were different. This semantics would dictate that we vary v to achieve the variation in x_1. Then of course x_2 would vary as well. So it would be true that were x_1 to be different, x_2 would be. And that seems a perfectly reasonable claim for some purposes. But not of course if we wish to read singular causal facts from our counterfactuals. So Pearl could argue that his semantics for counterfactuals connects singular counterfactuals and singular causal claims in the right way. And his semantics needs modularity. So modularity is a universal feature wherever singular causal claims make sense.

Laying aside tangled questions about the relations between singular causal claims and causal laws, which are the topic of this chapter, I still do not think this argument will work.

We could admit that for an epistemically convenient system Pearl's semantics for counterfactuals plus the counterfactual-causal links he lays out will give correct judgements about causal claims. We could in addition admit that causal claims cannot be judged by this method if the system is not epistemically convenient. All this shows is that methods that are really good for making judgements in one kind of system need not work in another kind.

More strongly, we could perhaps somehow become convinced that no formal semantics for causal claims that works, as Pearl's does, by transforming a test

---

[29] For Pearl this would mean that there was no alternative causal system possible that substituted the law 'let $x_1 = X_1$' for the law connecting x_1 and v.

into a truth condition, will succeed across all systems of laws that are not epistemically convenient. That would not show that there are no causal laws in those systems, but merely that facts about causal laws are not reducible to facts about the outcomes of tests.[30]

7.9   Conclusion

Modularity is not a universal feature of deterministic causal systems, nice as it would be were it universal. Part of my argument for this conclusion depends on asking of various factors, such as the fact that the toaster rack is bolted to the lever, 'Are these really causes?' I argued that they are not because they cannot do for us what we want these particular kinds of causes to do. In this case what we want is a guarantee that if we use these factors in applying the method of concomitant variation, the results will be reliable.

I think this is the right way to answer the question. We should not sit and dispute whether a certain factor in a given situation is really a cause, or what causality really is. Rather we should look to whether the factor will serve the purposes for which we need a 'cause' on this occasion. That means, however, that for different purposes the very same factor functioning in the very same way in the very same context will sometimes be a cause and sometimes not.

That is all to the good. Causality is a loose cluster concept. We can say causes bring about their effects, but there are a thousand and one different roles one factor can play in bringing about another. Some may be fairly standard from case to case; others, peculiar to specific structures in specific situations. Causal judgements, and the methods for making them reliably, depend on the use to which the judgement will be put. I would not, of course, want to deny that there may be some ranges of cases and some ranges of circumstances where a single off-the-shelf concept of causality, or a single off-the-shelf method, will suffice. But even then, before we invest heavily in any consequences of our judgements, we need strong reassurance both that this claim is true for the ranges supposed and that our case sits squarely within those ranges.

That of course makes life far more difficult than a once-and-for-all judgement, a multipurpose tool that can be carried around from case to case, a tool that needs little knowledge of the local scene or the local needs to apply. But it would be foolhardy to suppose that the easy tool or the cheap tool or the tool we happen to have at hand must be the reliable tool.

---

[30] For more about Pearl and counterfactuals, see ch. 10 in this book.

8 Against modularity, the causal Markov condition and any link between the two: comments on Hausman and Woodward

8.1 Introduction

This chapter is a commentary on a rich and intricate piece, 'Independence, Invariance and the Causal Markov Condition' by Daniel M. Hausman and James Woodward, which appeared in the *British Journal for the Philosophy of Science* in 1999. In this single article Hausman and Woodward defend two distinct theses with entirely separable arguments. Both theses are about causality and both are important.

The first thesis is that equations representing functional relations will be true causal laws if they are invariant under interventions into the independent variables. (They call this level invariance.) I found Hausman and Woodward's discussion of this thesis confusing. The primary reason, it turns out, is that they use different senses of 'intervention' and different senses of 'invariance' in the paper and they sometimes shift between them without saying so, probably because the two authors come to their joint work with different paradigms. But their thesis is true, and it is of considerable practical import since it provides a method for testing causal hypotheses. In section 8.2 I shall remark on these various uses as a help to other readers; and I shall back up Hausman and Woodward's claim by pointing to proofs that show that it is true under various different formulations, so long as the concepts employed line up in the right way.[1]

The second thesis has two parts. (1) Our concept of causal law implies that each separate effect in a set of causal laws must be independently manipulable. (They call this manipulability.) (2) For causal systems meeting a specific set of assumptions, thesis (1) (manipulability) implies the causal Markov condition. These, too, are important claims. The first puts a significant constraint on the kinds of system governed by causal laws. Together they justify the causal Markov condition, and that matters because we are then able to employ powerful

I would like to thank James Woodward and Daniel Hausman for setting me off on this work and for extensive correspondence. Research for this paper was carried out under a grant from the Latsis Foundation, for which I am extremely grateful.
[1] For proofs see ch. 10 in this book.

Bayes-nets methods for causal and probabilistic inference developed by Judea
Pearl and his associates and by Clark Glymour and Peter Spirtes and their associates.[2] Even without (1), thesis (2) is an exciting one, of great utility if true.
For it would mean that, if we have reason to believe that a system under study
satisfies the specified assumptions, then finding out that the laws of the system
are separately manipulable can be taken as a guarantee that the causal Markov
condition holds and thus that we are entitled to the use of Bayes-nets methods
for that system.

Both (1) and (2) are thus worth considerable attention. Unfortunately I do not
think either is true. With respect to (1), I shall argue that in the formulation that
Hausman and Woodward give it is patently false and that in a modified version
that is all they need for many of their purposes here and elsewhere it is still
false, though not patently so. Regarding (2), I shall argue that their proof for it in
the deterministic case is valid but vacuous: given the background assumptions
they make for the proof, the causal Markov condition holds whether or not
thesis (1) is true. So the hope that, once we know we are studying a system
of the right kind, separate manipulability will give us an independent test for
the causal Markov condition is dashed. For knowing that the system is the
right kind already guarantees the condition whether or not there is independent
manipulability; and that is a result that is well known. The proof in the purely
probabilistic case is invalid, I shall argue.

Hausman and Woodward also devote a number of pages to discussion of
earlier objections of mine to the causal Markov condition. I shall here argue that
the objections stand in the face of Hausman and Woodward's criticisms. Since
I have written at length elsewhere against both the thesis that effects must be
separately manipulable (see ch. 7 in this book) and against the causal Markov
condition,[3] I shall confine myself here to a discussion of the specific claims
that Hausman and Woodward make. I should note at the start, however, that
there is nothing incompatible between separate manipulability of each effect
and violations of the causal Markov condition, and the probabilistic models
I discuss that violate the causal Markov condition all allow each effect to be
manipulated on its own.

(Throughout I shall use capital letters for variables; lower case for values of
the variables. I shall also adopt the conventional notation where $X \perp Y/Z =_{df}$
$Pr(X = x$ and $Y = y/Z = z) = Pr(X = x/Z = z)Pr(Y = y/Z = z)$, for all
x, y, z. I use $Xc{\rightarrow}Y$ to stand for the causal law 'X causes Y' and I represent
causal laws that have the form of equations thus: '$X\ c^= f(\ldots)$', where the effect
appears on the left and the causes on the right. I will assume, as is usual, that
$X\ c^= f(\ldots) \rightarrow X = f(\ldots)$ but not the reverse).

[2] Pearl and Verma (1991); Spirtes, Glymour and Scheines (1993). For a discussion of Bayes nets
see ch. 6 of this book.
[3] See Cartwright (1999), ch. 5, Cartwright (2000) and ch. 6 of this book.

Also, when Hausman and Woodward discuss causal laws they generally have in mind a notion that is relative to a designated variable set: causal laws that cite direct causes relative to that variable set. I am more interested in the causal laws themselves; thus I will count as 'correct' a causal-law claim that represents correctly a true causal law even if the causes cited in the law are not direct causes relative to the variable set under consideration.

8.2   Intervention, invariance and modularity

Central to Hausman and Woodward's arguments for the causal Markov condition is their claim that the laws (if any) governing a set of factors **V** are not properly counted causal laws unless each of the factors represented can be manipulated separately, leaving the laws governing all the other factors in **V** intact. They call this feature of a causal system 'modularity' and express their claim thus:[4]

MOD: For all subsets **Z** of the variable set **V**, there is some non-empty range **R** of values of members of **Z** such that if one [HW-]intervenes and sets the value of the members of **Z** within **R**, then all the equations [that correctly record causal laws] except those with a member of **Z** as a dependent variable (if there is one) remain invariant. (Hausman and Woodward [1999], p. 545)

Hausman and Woodward have defined 'intervention' slightly differently from each other in other places. Here they do not offer a formal definition but tell us the intent:[5]

HW-intervention: An intervention, *I*, that sets the value of X 'is not an effect of any variable in **V**, *I* does not cause any variable in **V** by a path that does not go through X first, and *I* is not caused by any variable that causes any other variable in **V** by a path that does not go through *I* and X first'.

I call this 'HW-intervention' because I think it is not an adequate characterization of 'intervention'. In particular it is not enough to ensure either MOD or the result expressed in their first thesis that (with some caveats) a (true) functional law will remain invariant under interventions on the independent variables if and only if the functional law is also a true causal law. Their central idea is that 'causes are as it were levers that can be used to manipulate their effects'.[6] For a true causal law, when the cause changes (*ceteris paribus*), the effect should change in accord with the law – and the law itself should remain functionally true. (They call this level invariance.)

What follows the dash is the core of their first thesis, which was summarized briefly in section 8.1. Hausman and Woodward describe this thesis and defend it

---

[4] Hausman and Woodward (1999), p. 545.   [5] Hausman and Woodward (1999), p. 535.
[6] Hausman and Woodward (1999), p. 533.

with examples, but they do not state it formally. To state exactly what they mean takes some work, because at different points they offer different accounts of all three of the central notions involved in the claim: the notion of intervention, the notion of invariance and the assumptions about what it is for a causal-law claim to be correct; and they sometimes move around among these concepts without clear markers.

For example, they shift in the middle of the discussion of level invariance and modularity from one way of rendering the idea of an intervention – change the causal laws that govern the effect – to another – set the effect where we wish by manipulating one cause to offset the value induced by the other causes. Or they rather surprisingly tell us at one point to 'ignore the error terms'[7] in looking to see if a set of equations in which error terms appear are correct. They can do this because at this point they are supposing that correctness involves what I call 'prediction of first differences' (see ch. 10 of this book) instead of their earlier, more natural, sense that an equation is correct when it is literally true.

This could lead to confusion and to false results since what is true for one account of intervention will not be true for another and similarly with the other two concepts (see section 10.2). But the concepts can be defined formally and lined up in the right way so that Hausman and Woodward's thesis can be proved to hold for many kinds of causal systems (again see ch. 10). As remarked in section 8.1, the claim matters because it provides a useful test for causal laws in the kinds of systems where it holds.

The proofs, however, require an addition to the characterization of 'intervention' given above, and that is the case whether we have in mind intervention by changing the laws governing the effect or intervention by manipulating a special cause. The addition is just the requirement pointed to in MOD. When we intervene on X to check X's effect on Y, we must not only avoid affecting any other causes of Y or affecting Y directly; we must also avoid changing any of the causal laws that have Y as effects, except those connecting X's causes with Y via X. For instance, we must not change the very causal law that connects (or not) X and Y – either by adding a causal link that was not there or by destroying one that was.

If we are allowed to change the causal laws during an intervention, anything can happen. Consider a simple system with two causal laws: $L_1 : X \, \mathrm{c}^= aZ$ and $L_2 : Y \, \mathrm{c}^= bZ$. From these follow $L_3 : Y = bX/a$, which is thus functionally true though not a true causal law. Now let us perform our test. HW-intervene by changing L_1 to $L_1' : X \, \mathrm{c}^= I$, where I can take any value (in some appropriate range) that we wish; and do nothing else. L_3 is not true in this new system of causal laws; it is, as Hausman and Woodward teach, not invariant. But

---

[7] Hausman and Woodward (1999), p. 543.

look at a second case. HW-intervene by changing L_1 to L_1' and at the same time change L_2 to L_2' : $Y\, c^= bX/a$. This is clearly allowed by the definition of HW-intervention. Now L_3 is invariant. But it is not a correct causal law in the original system. The perfectly good test for whether $Y\, c^= bX/a$ is correct in the original system fails because we are too permissive in what we count as an intervention.[8]

MOD claims that cases like this cannot happen. Situations that satisfy the definition of an HW-intervention will never be ones in which more than the targeted law is changed. But that is not ruled out by the characterization of 'HW-intervention', so what is to stop it? One may feel queasy about the idea of changing causal laws at all. But if we can change one, why not two? It is clearly possible to have two causal systems that are identical in all but two, or three, or more of their causal laws. MOD tells us that this cannot happen if one of the laws in the second system simply fixes the value of one of the variables. But that is contrary to our experience.

The solution is to build the condition expressed in MOD into the definition of 'intervention'.

Intervention: I is an intervention on X if I is an HW-intervention on X and all causal laws stay the same except those that have X as effect or that have causes of X as cause and effects of X as effect.[9]

Hausman and Woodward are, I suppose, reluctant to do this because it weakens the usefulness of the invariance claim. We want to test a hypothesized causal law connecting X and Y by manipulating X. Our results do not tell us much unless we are assured that when we manipulate X we do not simultaneously change the causal laws of the system (except the ones noted), and that includes the causal law connecting X and Y. But that is what we wanted to find out about in the first place.

The situation is not so bad, however. For we may often be in a position to assume that what we do to change X has very little chance of changing the laws about what X causes even if we do not know exactly what those laws are. If we are not in that position, we are not able to rely on our test. The credibility of our results is dependent on the credibility of the assumption that we have succeeded in intervening properly.

If we adopt as the correct notion not 'HW-intervention' but 'intervention', MOD becomes trivially true and cannot do any work for us. But there is another

---

[8] A similar argument can be constructed for interventions that manipulate a special cause of X different from Z if there is one. But then the notions of correctness and invariance must be different as well.

[9] I am here allowing for transitivity of causal laws. For more precise definitions see ch. 10 of this book.

related thesis that might, and it is one that Hausman and Woodward's discussion clearly defends. I shall call it MOD#.

MOD#: for any variable set, **V**, it is always possible to intervene on each variable in **V**.

Why should we believe in MOD#? I have argued against it in ch. 7 of this book. Here I shall briefly discuss three central defences for it that Hausman and Woodward offer in their joint paper.

Their first argument supposes that 'if two mechanisms are genuinely distinct it ought to be possible (in principle) to interfere with one without changing the other'.[10] They then tell us, '[t]his understanding of distinctness of mechanisms plus the assumption that each equation expresses a distinct mechanism implies modularity: it is in principle possible to intervene and disrupt the relations expressed by each [causally correct] equation separately'.[11] (Note that here by 'modularity' it seems they mean MOD# and not MOD.)

It is the assumption that I quarrel with. The equations in question already have a job to do. The normal understanding is that we are discussing equations that (1) pick out for the given effect a full non-redundant set of causes:[12] and (2) lay out the functional form of the (true) causal law that holds between these causes and the effect. We can if we want change the subject. We can talk instead about sets of equations that represent relations each of which can be interfered with separately. But there is no reason to think that equations (if there are any)[13] that do this new job will have the characteristics usually connected with sets of equations that do the original job. So we must be careful at every step not to import without defence any facts true of equations that do the old job or draw any consequences about the old type of equation from anything we establish about the new.

Their second argument is this:

A system of equations that lacks modularity will be difficult to interpret causally. Suppose, for example, that when one [HW-]intervenes to change the value of Y ... [in equation (1) $Y = aX + U$], equation (2) [$Z = bX + cY + V$] breaks down in such a way that the value of Z does not change ... In this case, equations (1) and (2) do not fully capture the causal relationships in the system we are trying to model.

Their final sentence, it seems to me, is not true. We have a job we want equations (1) and (2) to do – give a full non-redundant set of causes for Y and Z and set

---

[10] Hausman and Woodward (1999), p. 549.   [11] Hausman and Woodward (1999), p. 549.

[12] For an account of what this means, see Cartwright (1989), p. 549.

[13] I have not yet figured out how to represent separate mechanisms by separate equations. Look for instance at the probabilistic equations in section 8.6 below. There is one equation for each separate effect. As I understand it Hausman and Woodward think that the two effects studied in this example are not produced by distinct mechanisms but rather by the same mechanism. So we should have one equation, an equation for the mechanism, rather than two. But what is this equation? What, for instance, are the quantities to be equated?

out the true causal law holding between these causes and their effects. The equations are not supposed to give information about why they are the true causal equations for the situation, or about what causal equations might hold if they did not hold. Why on the occasion is it impossible to change one without changing the other? Such information may exist but it is no job of these equations to convey it.[14] Nor can we assume that both jobs together can be done by the same equation.

Their third argument is that 'modularity provides a natural explication of what it is for a variable to be a direct rather than an indirect cause'. They illustrate by contrasting two causal systems:

$$\text{system 1} \quad Y \text{ c}= aX + U \tag{1}$$

and

$$Z \text{ c}= bX + cY + V \tag{2}$$

versus

$$\text{system 2} \quad Y \text{ c}= aX + U \tag{1}$$

and

$$Z \text{ c}= bY/a + V - bU/a \tag{2*}$$

Relative to the variable set $\mathbf{V} = \{X, Y, Z, U, V\}$, in system 1, X is a direct cause of Y, whereas in system 2, X is only an indirect cause. Hausman and Woodward point out that this shows up when we think about what happens under interventions on Y. If system 1 is the true system, equation (2) will be invariant under this intervention but (2^*) will not be; and conversely if system 2 is the true system.

Their discussion of these two systems points to an important result. In ch. 10 of this book I provide a proof of two theorems to the effect that, as Hausman and Woodward maintain, an equation is invariant under interventions into the independent variables if and only if it is causally correct.[15] Trivial lemmas to those theorems show essentially that an equation is not only causally correct

---

[14] The discussion becomes complicated because there are two ways to intervene: not only by changing the laws (1) and (2) but by manipulating the variables U and V. In this case Hausman and Woodward's idea is that, if there are cross restraints between U and V there must be a reason; and at base the only reason is that U and V have a common cause. Suppose I agree with them for the nonce. That goes no way towards showing that any two effects we may be concerned about must each have at least one cause they do not share in common. (I do not think they would disagree with this since Hausman, at least, offers independent reasons for thinking that this last is true. For my objections to these see ch. 7 of this book.)

[15] Hausman and Woodward (1999), p. 538. Note that there are some caveats to the two theorems proved there. See the proofs themselves for details.

but all of its causes are direct causes relative to **V** just in case the equation is invariant under interventions into each variable in **V**.

So there is a clear link between Hausman and Woodward's ideas about invariance and direct causation. But what about modularity? Before turning to that, let me take up another issue. For I think the importance of Hausman and Woodward's discussion here goes well beyond questions about direct causation, which are relative to the choice of variable set, to questions about the causal laws themselves

Under plausible assumptions about transitivity and consistency in causal systems (see ch. 10 of this book), systems 1 and 2 are incompatible; both could not be true of any single situation. Hausman and Woodward explain this way: 'From the point of view of (1)–(2), (2*) entangles distinct mechanisms ... one of which links X to Y and one of which links X and Y to Z'.[16] We can make the same point without escalating to talk of mechanisms and entanglement but using only the vocabulary of causal laws, which is already presupposed in talking of mechanisms: (2*) is functionally true as a consequence of two distinct causal laws but is not itself a true causal law.

What though does it mean to say that (assuming system 1 is correct) (2*) does not represent a true causal law? We might be inclined to say that a causal-law claim is true if all the independent variables are genuine causes of the targeted effect and the law is functionally true. But that will not do, as we see from looking at (2*), which satisfies both these criteria. It seems we must give the answer that I gave above: the law claim must present nature's true functional form for the law. But what does it mean to say that (2) gives nature's true coefficient for Y's influence on Z and (2*) does not?

This is an important question and Hausman and Woodward's discussion offers a partial answer that is non-inflationary.[17] As I remarked, we have theorems that show that an equation like (2*) is correct if and only if it is invariant under interventions on right-hand-side variables. This gives a kind of operationalization for the idea that a functional form for a causal law is nature's true one. So the invariance results do not just provide a test for correctness of functional form, but in providing a test they help give sense to the concept. We still, however, face the question I earlier laid aside: this makes a connection with Hausman and Woodward's concept of invariance but where does modularity enter?

---

[16] Hausman and Woodward (1999), p. 543.

[17] My own answer presupposes a notion of causality stronger than that of 'causal law' – the notion of a capacity. The coefficients in equation (2) represent the strength of Y's capacity to cause Z. Although this is an inflationary answer, in *Nature's Capacities and their Measurement* (1989), I, like Hausman and Woodward here, offer various procedures for operationally determining what this strength is – that is, for measuring the capacity. Unlike the supposition of Hausman and Woodward, I do not, however, think it necessary that the concept be measurable each and every time it applies in order to be legitimate (see discussion in ch. 7 of this book).

Because I have been describing theorems rather than writing them down, I have been sloppy about a number of aspects of the formulation, one of which matters here. The theorems are to the effect that (with caveats not relevant here) for a given causal equation, if it is possible to intervene in the right-hand-side variables of that equation, the equation is causally correct iff it is invariant under those interventions. Modularity guarantees that such interventions are always possible. (Note that again it is MOD# that matters and not MOD.)

Do we then need modularity? That depends on how strong an operationalist one wants to be. P. W. Bridgman is the arch operationalist.[18] He insisted that a concept did not make sense unless it could be operationalized and that that operational procedure could be used to test for it wherever it was supposed to apply. We might, for example, operationalize the concept of length using a foot ruler; we could not then sensibly talk of the size of a molecule.

Hausman and Woodward have found a way to measure for correctness of causal laws (including the functional form of the causal laws). They insist that the test must be applicable wherever the concept of causal law applies. I disagree. We have many concepts that do not reduce to others supposed to be more readily observable. 'Size' is one example, if we intend to include molecules in its extension; 'causal law' is another, I maintain. We often have good reasons for postulating these concepts and can say a lot about them; and in many situations we can measure them in a particular way – that is we have necessary and sufficient conditions for their correct application. I think that is enough.

Hausman and Woodward's position is far too strong. In the first place it overlooks the possibility of devising other methods of measuring the concept in circumstances where the first method cannot be used, or other indirect methods of testing, like the hypothetico-deductive method. But besides that, their requirement is too strong in itself. The fact that we can provide a test that, so far as we can tell from independent evidence, gives the right results wherever we can apply it, supports our belief in the concept. We should not, then, withhold the concept from situations that seem the same in all other ways relevant to its application just because our test cannot be applied in those situations.

Bottom line on invariance and modularity: Hausman and Woodward's concept of invariance provides a powerful tool for testing for correctness of causal laws, including their functional form. But their views on modularity seem to have problems. First, the formulation they give for their claim, MOD, seems trivially false. Second, another related view they defend – MOD# – though not trivially false is far too strong.

---

[18] Bridgman (1927).

8.3   The causal Markov condition: CM1 and CM2

The causal Markov condition is defined relative to a set, **V**, of random variables:

CM: For all distinct variables X and Y in the variable set **V**, if ¬(X c→Y) then X⊥Y/Par(X).[19]

Here '*Par(X)* is the subset of **V** containing all the direct causes of X in **V**'[20] (p. 524).

Hausman and Woodward divide this into two claims:[21]

CM1: If ¬(X⊥Y) then X c→Y or Y c→X or X and Y are effects of some common cause.[22]

CM2: ∀X, Y in **V** for which ¬(X c→Y), if Par(X) is nonempty then X⊥Y/ParX.

They remark, 'one might accept CM1 but deny that causes always screen off in the manner required by CM2. This last position . . . may be Cartwright's'.[23] It is in fact not my position. For some kinds of systems of causal laws, CM1 can be proved; for instance, for a 'triangular' system of deterministic causal laws of the form: $X_n \ c = a_1 X_1 + \cdots + a_{n-1} X_{n-1} + a_n U_n$, where a probability measure is specified over $(U_1, \ldots, U_n)$ under which the U's are mutually independent in all combinations. (This fact is well known, but for illustration, a proof is provided in the appendix.) Should we expect something like CM1 to hold for every system of laws, causal or non-causal, deterministic or probabilistic? I do not see any reason to think so.

Ideally what I would like to see are formal presentations of different kinds of systems of laws for which we can prove whether an analogue of CM1 holds or not. This gives us clear information to help in our decision about whether to assume CM1 for any particular case we want to treat. If we are prepared to model the laws at work in that situation with one or another of our formal systems, we will then know whether we can or cannot avail ourselves of CM1.

About CM2 I can be more specific. For deterministic causation, if our variable set is complete enough, CM2 is trivially true. The only questions that remain concern what kinds of minimal conditions we can put on the variable set to ensure CM2. We will see one such set of conditions in section 8.4.

For probabilistic causation the situation is different. Whenever a deterministic cause obtains, all of its effects obtain on every occasion. For an indeterministic cause, the cause may be present and one or another or all of its effects fail to obtain. We often approach the question of probabilistic causality from the point

---

[19] Hausman and Woodward (1999), p. 523.
[20] Hausman and Woodward (1999), p. 524.   [21] Hausman and Woodward (1999), p. 524.
[22] Even though the argument given in footnote 5 (p. 524) does not entirely work to establish that CM implies CM1, it is true.
[23] Hausman and Woodward (1999), p. 524.

of view of the effect: we are interested in the effect and how to obtain or prevent it. This naturally leads to a focus on the partial conditional probability for the effect to occur given the cause. But from the point of view of the cause, a joint conditional probability over the entire outcome space of its effects is required to model its behaviour. Total independence between the effects is just one very special way the effects can relate.

If we focus on a simple case with yes–no variables, we can see how special the case of independence between the effects is. Given a cause C and effects X and Y, $X \perp Y/C$ if something we might call the causal Markov constraint (CMC) is satisfied:

$$\text{CMC}: Pr(+x + y/ + c)Pr(-x - y/ + c)$$
$$= Pr(+x - y/ + c)Pr(-x + y/ + c)$$

Nothing in the concept of causation suggests that CMC should be satisfied or that in general all the effects of a given cause should be independent. Causes make their effects happen; so *ceteris paribus* a cause will increase the probability of its effect. But that leaves open the question of how the production of one effect bears on another. Independence fails whenever effects are produced in tandem. This means that on any occasion on which a probabilistic cause produces both effects and side effects, there will be violations of independence.

To all appearances these occasions are widespread. In criticizing the causal Markov condition in earlier work,[24] I have discussed the example of a chemical factory, C, that causes a nasty pollutant Y as a side effect whenever it produces X, a chemical used for treating sewage. The factory is a purely probabilistic cause of both effects: the chemical and the pollutant are always produced together but they occur only 80 per cent of the time the production process is in place. I shall return to this example in sections 8.6 and 8.8 since Hausman and Woodward take it up. But this is just one made-up example among a host of real-life cases that we regularly model this way: viruses often produce all of a set of symptoms or none; we raise the interest rate to encourage savings, and a drop in the rate of consumption results as well; we offer incentives for single mothers to take jobs and the pressure on nursery places rises; and so forth.

Hausman and Woodward, like many others, propose that examples like this often involve a too-coarse-grained description of the cause – underneath the processes are deterministic. Perhaps so, perhaps not. We have trouble getting models at all. Our best models are often probabilistic. I would urge against the assumption that all of the untidy world outside of quantum mechanics is deterministic. But in a sense that is beside the point in discussing the paper of Hausman and Woodward. If one wants to insist on determinism, or to forbid

---

[24] Cartwright (1999; 2000).

probabilistic causes to produce side effects, the issue is settled: we do not need a lengthy discussion to impose the causal Markov condition.

I said above that nothing in the concept of causation forces a cause to produce each of its distinct effects separately from one another. Hausman and Woodward think they have found something – modularity. As we have seen, they take modularity to be a condition that a system of laws must satisfy before it counts as causal. Their contention is that modularity (or a related condition, the 'strong independence assumption' that Hausman[25] has argued to be a prerequisite for counting relations as causal) ensures the causal Markov condition. They offer three proofs, two for deterministic causality and one for probabilistic. I shall discuss each in turn.

8.4   From MOD to the causal Markov condition and back

Hausman and Woodward argue that MOD implies a principle called MOD^* and that MOD^* holds if and only if CM2. This is the exciting claim I described in section 8.1. Their proof for it proceeds in three stages. Stage one aims to establish

$$\text{MOD} \rightarrow \text{MOD}^* \tag{1}$$

where

$$\text{MOD}^*: \forall X, Y, \text{set-}X, (X \text{ is distinct from } Y) \ \& \ \neg(Xc \rightarrow Y) \rightarrow Y \perp \text{set-}X$$

and set-X is a random variable that represents an HW-intervention on X to set it at any prescribed value. Stage two aims to establish

$$(=) : Y \perp \text{set-}X \text{ iff } Y \perp X/Par(X)$$

and thus to show

$$\text{MOD}^* \text{iff CM2} \tag{2}$$

and thereby establish the claim that I am primarily concerned about,

$$\text{MOD} \rightarrow \text{CM2} \tag{3}$$

Stage three interprets these results to mean that 'the independent disruptibility of each mechanism turns out to be the flip side of [CM2]'.[26]

I find problems at each stage of the proof. The first may be just a difficulty I have in interpreting what Hausman and Woodward write, for there is a valid proof of (1) along the lines they lay out. At the second stage, the argument is invalid, but the overall effect can still be achieved with the addition of an extra premise, and in more than one way.

---

[25] Hausman (1998).   [26] Hausman and Woodward (1999), p. 553.

But the problem at the third stage, I think, will not go away: their interpretation of the results is misleading. Both CM2 and MOD# (my version of MOD that is not patently false) are separately derivable from the background assumptions alone; neither has any special role in the derivation of the other. Given the background assumptions Hausman and Woodward make in order to prove that MOD# implies CM2, CM2 holds anyway, and it continues to do so even if we drop some assumptions crucial for ensuring MOD# itself.

Hausman and Woodward put three conditions on the variable set, **V**, under study. These are (i) determinism; (ii) additivity of missing causes and (iii) causal sufficiency, which I formulate thus:

(i) + (ii) $\forall V_i$ in **V** (V_i c$=$ $f_i(Par(V_i)) + a_i U_i$), where a_i is either 0 or 1;

(iii) $\forall X$, Y in **V** (Z is a common cause of X, Y → Z is in **V** unless there are other common causes of X and Y in **V** such that Z's influence on X and Y is entirely via these other common causes);

They assume in addition:

(iv) CM1;

(v) HW-interventions can be treated as random variables (which will be labelled 'set-V' for V in **V**);

(vi) 'the existence of unrepresented causes': $\forall V_i$ in **V**, $a_i = 1$.

Stage one MOD → MOD*. Paragraph two of Hausman and Woodward's argument for MOD* says, 'since [set-X] is not an effect of any variable in **V** and is causally related to variables in **V** only by virtue of being a direct cause of X, **CM1** implies that *it is probabilistically independent of all other interventions and of everything that is not an effect of X*'.[27] The part I have italicized is the consequent of the claim they wish to establish. I agree that what they say is correct: CM1 and causal sufficiency imply that ¬(X c → Y) → set-X ⊥ Y. But where does MOD enter? Just after this paragraph Hausman and Woodward claim, given CM1 and (v), 'MOD thus implies by the argument given immediately above MOD*'.[28] It looks from their own argument in this paragraph as if MOD is completely irrelevant. Given the premises it does imply MOD* but just because given the premises MOD* is true anyway.

What role then does MOD play? Well, here is a worry that MOD might answer. If the causal laws relating to Y are different when set-X has one value from when it has another, $Pr(Y/X)$ can differ for different values of X even if ¬(Xc → Y) in either set of laws. Of course in this situation CM1 need not be true either, even though it may hold in each set of laws separately. The reason that the role of MOD in the proof of MOD* is not clear is that CM1 itself is

[27] Hausman and Woodward (1999), p. 552.   [28] Hausman and Woodward (1999), p. 553.

expressed in shorthand. Written out more fully, read CM1 (long version): for any additive[29] deterministic causal system $\langle E, \Lambda, Pr \rangle$ and for any causally sufficient sets $\mathbf{V}$, of variables representing E or any subset of E, $\forall X, Y \in \mathbf{V}, Pr\,(Y/X) \neq Pr\,(Y) \rightarrow X\,c \rightarrow Y$ or $Y\,c \rightarrow X$ or $\exists V \in \mathbf{V}(Vc \rightarrow X$ or $Vc \rightarrow Y)$, where E is an ordered n-tuple of effects, Λ is a set of causal laws for those effects of the form $E_i\,c = f_i(V_1^i \ldots V_j^i) + a_i U_i(V_j^i \in \mathbf{V})$, and Pr is a probability measure over $E \cup \{U_1, ..., U_n\}$. Given the long version of CM1, we can now see how MOD comes into play in the proof of MOD*: we need it in order for CM1 to apply, since CM1 refers only to a single causal system and not to a patchwork over many. Perhaps this is what Hausman and Woodward intended by the first paragraph of their argument for MOD*. At any rate, it does show that their claim is not vacuous that given CM1 (and other background assumptions), MOD implies MOD*.

Bottom line for stage one: there is a valid argument from MOD to MOD* along Hausman and Woodward's lines, and it uses MOD in an essential way.

Stage two   The first part of the argument for $(=)$ is unproblematic: CM1 is used to establish that $\forall X$ in $\mathbf{V}$, $U_x \perp Par(X)$. Next we are told, '[s]ince, conditional on $Par(X)$, the only source of variation in X is U_X, any variable Y in $\mathbf{V}$ distinct from X can covary with X conditional on $Par(X)$ iff it covaries (unconditionally) with U_X [. . .]. So $Pr(X/Par(X)\&Y) = Pr(X/Par(X))$ iff $Y \perp U_X$. Last, '[s]ince U_X satisfies the definition of an intervention, one can infer [. . .]$(=)$'.[30]

The trouble is that the second claim is false, and for exactly the reason they highlight in parentheses. What is true is that Y can co-vary with X conditional on $Par(X)$ iff it co-varies with U_X, not unconditionally, but conditionally on $Par(X)$. So what we can establish is only $(=)_{\text{weak}}$: $Pr(\text{set-}X/Y\&Par(X)) = Pr(\text{set-}X/Par(X))$ iff $Pr(X/Par(X)\&Y) = Pr(X/Par(X))$. I suppose it is worth showing this formally since Hausman and Woodward make a point of claiming that unconditional covariation of Y with U_X is enough. We know $X = f_X(Par(X)) + U_X$. Then $Pr(X = x/Par(X) = p$ & $Y = y) = Pr(U_X = x - f_X(p)/Par(X) = p$ & $Y = y)$. Compare $Pr(X = x/Par(X) = p) = Pr(U_X = x - f_X(p)/Par(X) = p)$.

Hausman and Woodward's claim $(=)$ follows only if $Pr(U_X/Par(X)\&Y) = Pr(U_X/Par(X))$. But we have no guarantee of this. We do have $U_X \perp Par(X)$ but that does not ensure that $U_X \perp (Par(X)\&Y)$. Looking to the overall proof, we need only consider cases where $\neg(Xc \rightarrow Y)$, from which it should follow

---

[29] See Hausman and Woodward's footnotes 15 and 16. One reason for requiring additivity of the exogenous factors is to get this requirement right. Otherwise we could admit causal laws of the form $X_i\,c = \delta(U_i = u)f_i^u(X_1 \ldots X_{i-1}) + \alpha U_i$. CM1 (as well as various other conditions we commonly assume) will not hold for causal systems with laws of this form.

[30] Hausman and Woodward (1999), p. 554.

that $\neg(U_X c \rightarrow Y)$, and by CM1 and causal sufficiency, $U_X \perp Y$. But even that does not get us what we need since it is possible that $A \perp B$, $A \perp C$ and $B \perp C$, and yet $\neg(A \perp (B \& C))$.

There are two ways around the problem that I can see. Both require premises additional to those assumed by Hausman and Woodward. The first is to insist on stronger conditions on the U's. This is what Pearl and Verma, and Spirtes et al. do:[31] we can restrict our results to causal systems in which the U's are not just pairwise independent but are independent in all combinations.[32] These are the 'strong independence assumptions' that characterize 'pseudo-indeterministic systems', which Hausman and Woodward refer to in their immediately following section.[33]

With regard to these stronger independence assumptions, I suppose we could try to revise CM1 so that it allows us to infer not just that $U_X \perp U_V$ for all V but that U_X is orthogonal to the whole set of other U's in all combinations. I myself do not know how to do this without losing the intuitive basis for CM1, since I do not know how to think about causal laws with respect to arbitrary combinations of effects. (For instance, does $Xc \rightarrow Y$ imply either $Xc \rightarrow Y \& Z$ or $Xc \rightarrow Y \vee Z$?)

The second way around the problem also introduces an additional premise. Since we will in the end only be concerned with Y such that $\neg(Xc \rightarrow Y)$, let us restrict the discussion to this case. From this we know that $\neg(U_X c \rightarrow Y)$. This means that it is not a causal law that U_X causes Y in the population under consideration, the population described by the probability measure Pr. Let us think about subpopulations of that population, those defined by $Par(X) = p$. I take the following to be true about causal laws:

(vii) *causal dilation*: if $Ac \rightarrow B$ in a population, Q, $Ac \rightarrow B$ in any larger
      population in which Q is totally included.

Given causal dilation we can conclude that $\neg(U_X c \rightarrow Y)$ in the subpopulation picked out by different values for $Par(X)$. We know further that Y does not cause U_X in the population under consideration and that they do not have a cause in common. So by causal dilation and CM1, we can infer that $U_X \perp Y / Par(X)$.

Should we believe in causal dilation? Many authors writing about generic causal laws do not provide an analysis of them, although they often provide

---

[31] Pearl and Verma (1991) and Spirtes et al. (1993). I take it they, but not I, think this is expressed in MOD.

[32] Or, alternatively, that each U_n is independent of all $V_1, \ldots, V_{n-1}$ in $\mathbf{V}$ in all combinations.

[33] Even there, however, they do not note the necessity that the U's be independent in all combinations. In fact, they repeat exactly the argument that they give in the previous section, making us wonder why they call these 'strong' independence assumptions. Note also that this condition on the U's is not Hausman's own condition (which I discuss in section 10.6), which they also call 'strong independence'.

partial implicit definitions by listing constraints that they suppose systems of causal laws must satisfy. My own work begins from the notion of singular causation, which I take to be primitive. Then I can give truth conditions for generic causal laws:[34]

$Ac \rightarrow B$ is a true generic-level claim about a population iff in that population it is reliable that some A's cause B's.[35]

Given this account of causal laws, causal dilation will clearly be true.

This reading is the one adopted by Spirtes, Glymour and Scheines at the beginning of their book; and it is one that both Hausman and Woodward should be sympathetic with too. But it should be pointed out that it is not an account that allows for a close parallel between correlation and causation in a given population. Even if we suppose that effects are probabilistically dependent on their causes in the 'right kind' of populations, Simpson's-paradox reversals allow that they may be independent in larger populations of which these are a part.[36] The reading also allows that the same factor may both cause and prevent a given effect in the same population. I think that this is as it should be. My argument for $U_X \perp Y / Par(X)$ when $\neg(Xc \rightarrow Y)$ will not work for those who think otherwise.

It is trivial now to complete the proof of the derivation of CM2 from MOD*. We have established that, given (i) to (vi), if $\neg(Xc \rightarrow Y)$, $Pr(X/Par(X)\&Y) = Pr(X/Par(X))$. So we have arrived at CM2. I shall discuss the reverse inference, from CM2 to MOD*, at the end of this section.

Bottom line for stage two: we can establish MOD* $\rightarrow$ CM2 if we add further plausible premises to those of Hausman and Woodward.

Stage three What then of Hausman and Woodward's claim that 'the independent disruptibility of each mechanism turns out to be the flip side'[37] of CM2? Note first that Hausman and Woodward do not claim to argue from CM2 to MOD, but only from CM2 to MOD*. Nor should we expect to be able to go readily from MOD*, which is about probabilistic independencies, to MOD, which is about the invariance of equations. And of course we cannot get to MOD from MOD* with the additional (correct) claim that Hausman and Woodward make that, given our background assumptions, MOD implies MOD*! By their

---

[34] Cartwright (1989).

[35] We are dealing here with probabilistic causation. 'Reliable' is a fudge word to deal with the fact that the claim may fail in finite populations, especially if they are small. The point is that some A's will cause B's in the targeted kind of population 'in the long run'.

[36] Spirtes et al. (1993) try to 'have their cake and eat it too' in this regard. They adopt the truth condition I propose but still require B to be conditionally dependent on A in any population in which '$A c \rightarrow B$' is true by maintaining that Simpson's-paradox reversals will 'almost never' happen.

[37] Hausman and Woodward (1999), p. 553.

own report, what they establish is 'the equivalence of MOD* and CM2'[38] and not the equivalence of MOD and CM2.

If we are concerned to ensure that MOD obtains, perhaps we do not need to worry though. For assumptions (i)–(vi) may seem to imply MOD directly. Indeed, just (i)–(iii) may seem to suffice. For it looks as if U_X can be changed arbitrarily while all the other U_v stay the same, thereby providing way to set X, and all the equations will remain invariant. But that does not provide an argument that takes us from (i)–(vi) to MOD, for two reasons.

First, we need to add the requirement that there are no cross-restraints among the values of the U's. This does not follow from any of the conditions presupposed. The second is the reason I have mentioned in section 8.2. I do not see how to get MOD without building it into the definition of 'intervention'. MOD says that *every* HW-intervention leaves the equation invariant, but the U's only provide *some* HW-intervention that leaves equations invariant.

If we add the further assumption:
(viii) there are no cross-restraints among the values of the U's,
we can at least establish MOD#, though not MOD.

Given (i)–(viii) there is then a result we seem to be able to prove that may seem to support the Hausman and Woodward claim that modularity is the flip side of CM2:

MOD# iff CM2

But it is important to notice that the proof uses a very odd, though valid, method to establish the equivalence: we begin from the background assumptions and establish each side of the equivalence separately, without invoking the other. MOD# holds given the assumptions. So does CM2. So, given the background assumptions MOD# iff CM2. And it is even the case that premises that are essential for the one are not essential for the other and the reverse.

What really matters for MOD# are the assumptions that there are unrepresented causes and that their values are mutually unconstrained. These are what guarantee that we can intervene without changing any equations. But CM2 holds whether or not there are unrepresented causes.[39] (Just look at the formula I wrote down for the proof. If there are no U's the equivalence is trivial.)

What matters for CM2 is this: whatever unrepresented causes there are, they are probabilistically independent of each other. So we do not need MOD or MOD# for CM2; and even if we have it, the 'no cross-restraints' assumption

---

[38] Hausman and Woodward (1999), p. 554.

[39] There is one place in their paper where Hausman and Woodward seem to recognize this. On p. 555 they remark that it may seem odd that we need to assume the existence of unnecessary causes and explain that that is because 'the "only if" part of (=) requires that there be an unrepresented source of variation'. But just one sentence later they again maintain the connection: '**CM** follows from the view that causes can be used to manipulate their effects'.

on the values that guarantee MOD# is not enough to give us the probabilistic independence we need for CM2. CM2 is guaranteed by CM1, causal sufficiency and causal dilation; or alternatively by the stronger independence assumptions described. Assumptions about lack of constraints on values are not of help unless we are prepared to make some controversial extra-logical inference from lack of constraint on values to probabilistic independence.

Return then to the claim that CM2 and modularity are the flip sides of each other given the background assumptions. To interpret the results thus would be extremely misleading. For both claims – MOD# and CM2 (though not MOD itself, which I have argued is almost always false) – follow separately just from the background assumptions. Normally when we say that A is the flip side of B we mean that both A and $\neg A$ are possible and so too with B and $\neg B$, but we always have either A and B together or $\neg A$ and $\neg B$. The claim that Hausman and Woodward make is like saying where A and B are both true that A is the flip side of B. Or, given the previous two paragraphs, like saying A is the flip side of B given WXY because WX implies A and WY implies B.

For completeness we should look again at the equivalence that Hausman and Woodward explicitly claim to prove: MOD* iff CM2. As I remarked, showing this is a long way from showing MOD or MOD# iff CM2 since MOD* does not imply either MOD or MOD#. On the other hand, as we have seen, MOD does imply MOD*. If indeed MOD* implies CM2 then we seem, as I said in section 8.1, to have a very interesting result about when CM2 obtains. But the result is a chimera, for the reasons laid out in the paragraphs above: given the background assumptions, CM2 obtains anyway, whether or not MOD or MOD# or MOD* holds.

Let us first review MOD∗ → CM2. Suppose we deny MOD* (and thereby MOD). We still have CM2, given premises (i)–(vi) and causal dilation, and we can prove it by the same argument that Hausman and Woodward give.[40] The only place in their argument that MOD* appears to play a role is in the last full paragraph on p. 533, where they argue, 'U_X thus satisfies the definition of an intervention ... and [therefore] given CM1, ... U_X is probabilistically independent of Par(X)'. But it does not matter whether U_X satisfies the definition of an intervention or not. Given CM1, U_X is independent of $Par(X)$ anyway.

Going the other way from CM2 to MOD* is more of a problem. For here we cannot even establish the claim that CM2 → MOD* given our background assumptions by using the grounds I have been pointing to, that the background assumptions imply the consequent all by themselves. For without MOD, they do not do so.

The argument that I think Hausman and Woodward intended in this direction has a mistaken quantifier shift. Here is the argument in sketch. Assumption (vi)

---

[40] Assuming we fix it up in the ways I have suggested.

assures us that there is an unrepresented cause, U_v, for each V in **V** and (iii) and (iv) (causal sufficiency and CM1) assure us that each of these causes, U_v, will be independent of all factors except effects of V. So if $\neg(Xc \to Y)$, $U_X \perp Y$. But, as we have seen 'U_X ... satisfies the definition of an intervention with respect to the X'. Call the intervention 'set-X'. Then we have arrived at MOD*: if $\neg(Xc \to Y)$, set-$X \perp Y$.

The argument is not valid, however. For MOD* – at least so far as I have been reading it – says that all HW-interventions on a variable X are independent of everything except the effects of X; that is, MOD*: $\forall X, Y \in$ **V**, $\forall$ HW-interventions set-X, $(\neg(Xc \to Y) \to Y \perp$ set-$X)$. The argument shows only that there is some intervention on X that is independent of the effects of X. The argument I have sketched makes the simple mistake of inferring from 'U_X is *an* intervention on X' and '$U_X \perp Y$' to 'for *any* intervention on X, set-X, set-$X \perp Y$'. And I do not know of any other argument that will secure the conclusion from the seven premises assumed.

We can of course render their MOD* differently. As it is, MOD* follows from MOD. I have argued that MOD is false and have proposed to substitute MOD# for it. We can similarly formulate

(MOD#)*: $\forall X, Y \in V$, $\exists$ HW-intervention 'set-X' [set-X leaves all the eqtns of the sys invariant & (X is distinct from Y) and $\neg(Xc \to Y) \to Y \perp$ set-X)].

Now it is true that given our background assumptions, CM2 implies (MOD#)* in the Pickwickian sense that the background assumptions themselves imply (MOD#)*. The converse is true as well, so it is indeed the case that (MOD#)* iff CM2, given the assumptions. So there is a reading – (MOD#)* – under which Hausman and Woodward's claim of the equivalence of MOD* and CM2 is true. But I do not see what special significance we can attach to this result.

Bottom line for stage three: Even given (i)–(viii) Hausman and Woodward's strong claim of modularity can still be false, so we will not have MOD iff CM2. It is true that MOD# iff CM2, though in a very Pickwickian sense since the premises imply each side of the equivalence separately; so it would be misleading to say that CM2 is the 'flip side' of MOD#. Similarly, '(MOD#)* iff CM2 is true in the same Pickwickian sense. But it is not true even in this odd, weak sense that MOD* iff CM2, as Hausman and Woodward claim.

8.5   A second argument for CM2

In their section 8.8 Hausman and Woodward offer a second proof for the causal Markov condition. As with the first, I had some difficulty understanding the precise argument. I also had trouble understanding the exact ways in which it was supposed to differ from the first. What follows is my best interpretation; I hope my account of it is close to their intentions and intuitions.

The argument explicitly uses premises (i)–(iv) plus two new premises. The first they describe as 'a strong independence condition: Every variable X in **V** has some cause that is not in **V** and bears no causal relations to anything in **V** apart from those that result from its being a direct cause of X'.[41] As they note, 'strong independence' follows from (iv), (CM1) and (vi) (the existence of unrepresented causes). The second premise is 'the assumption that a conditional sample dependency operationalizes a counterfactual concerning what non-accidental connections would obtain if the condition were met'. Both new assumptions turn out to be irrelevant to the proof, if my understanding of the proof is right.

Here is their argument:

Suppose then that X and Y are probabilistically dependent conditional on Par(X) ... From the counterfactual operationalization, it follows that X and Y would be non-accidentally connected if all the parents of X were unchanging. So if all the parents of X were unchanging, then (by CM1) X and Y would be connected as cause and effect, or they would be connected as effects of a common cause[42] ... If all the represented direct causes of X were unchanging, then X and Y would not be effects of a common cause because the only source of X's variation, by the strong independence condition, bears no causal relation to Y except via causing X. Nor could Y cause X, because all the causal influences on X apart from its unrepresented cause have been frozen. So ... X cause Y.[43]

Hausman and Woodward point out that if there is a sample correlation between *X* and *Y* conditional on *Z*, we can use their second new premise plus CM1 to conclude that *X* and *Y* 'would be related as cause and effect or as effects of a common cause – that is that there would be a non-accidental connection between X and Y if the value of Z were observed to remain fixed at z'.[44] This last clause gives us an account of what they mean by non-accidental connection. The premise, so far as I can see, is irrelevant because no part of CM2, or anything else that enters as premises, have anything to do with sample dependency: the discussion is entirely about probabilistic dependency.

Next let us get rid of strong independence, which I think we do not need either. What strong independence adds to (i)–(iv) is (vi), the existence of unrepresented causes. Yet we know CM2 is trivially true for deterministic systems if there are no unrepresented causes. What we need then is a proof that shows what happens *if* there are unrepresented causes, and for this we do not need a premise that

---

[41] Hausman and Woodward (1999), p. 558.

[42] I leave out here and throughout the discussion of their section 8 their caveats that the correlations might be 'brute-fact' (Einstein–Podolsky–Rosen)-type correlations. That is because I do not see why we need this caveat given CM1.

[43] Hausman and Woodward (1999), p. 559.   [44] Hausman and Woodward (1999), p. 559.

says that there are unrepresented causes. All the work of strong independence in the proof is done by CM1, (iv) and causal sufficiency (iii).

Let us turn then to the proof for the case where there is an unrepresented cause for X and hence there is variation in the value of X when the parents of X are held fixed. Following Hausman and Woodward we suppose (for purposes of a *reductio*) that $\neg(Xc \to Y)$ and $\neg(X \perp Y/Par(X))$. From the discussion in section 8.4 above, we know this last obtains iff $\neg(U_X \perp Y/Par(X))$.[45] We can grant that $\neg(U_Xc \to Y)$ since $\neg(Xc \to Y)$ and that $\neg(Yc \to U_X)$ and that U_X and Y do not have a common cause, from causal sufficiency and CM1. That ensures that $(U_X \perp Y)$. But, as before, how do we get $(U_X \perp Y/Par(X))$?

Let us follow Hausman and Woodward's lead and look to subpopulations in which the parents of X have fixed values. Consider the causal laws that obtain in one of these subpopulations, S, and their relation to a new probability measure, $Pr'(\ldots) =_{df} Pr(\ldots /Par(X) = p)$. By hypothesis under Pr', $\neg(U_X \perp Y)$. So by CM1, U_X and Y must be related as cause and effect or must have a common cause under the causal laws that obtain in S. If we adopt, besides (i)–(iv), causal dilation as well, these alternatives will be ruled out, as in my argument in section 8.4 above. So by *reductio* the result is established.

Recall that the standard proofs of the causal Markov condition assume that unrepresented causes are mutually independent in all combinations. This proof allows us to weaken that assumption: we only need the U's to be pairwise independent – given causal dilation. If the argument I construct here is indeed the kind of thing Hausman and Woodward had in mind, it warrants note. I am not sure though if this is what they intended because, to get to the conclusion, we had to drop two premises and add one. The addition may be one they presupposed, but the premises deleted are ones they claim to matter.

8.6   The proof of the causal Markov condition for probabilistic causes

In their section 8.6 Hausman and Woodward claim 'CM holds in indeterministic circumstances',[46] and they offer two arguments for this conclusion. I think the conclusion defended, though, is that CM2, not CM, holds in indeterministic circumstances since in both arguments CM1 appears as a premise and never as a conclusion. I start by remarking that this claim is false. My chemical factory example is a clear counterexample. So how can they have proved it? Let us start with their second argument since it will be very familiar by now.

---

[45] Hausman and Woodward talk about the dependence of X and Y. I talk about the dependence of U_X and Y because if X and Y are not related as cause and effect, CM1 only requires that they have a common cause. It does not require that the common cause 'account' for variations in X. So the common cause could be among the parents of X.

[46] Hausman and Woodward (1999), p. 570.

Hausman and Woodward report, '[t]he second [argument] retraces essentially the argument given above in section 8.7 [i.e. the argument discussed in section 8.6 here] in a form that is appropriate for indeterministic causal relations' (p. 570). Indeed I think it does, and in so doing it suffers from the same defects as those I pointed out in section 8.4 – with one addition: this time the overall argument is invalid and the conclusion, I think, is not just vacuous but false.

As in the deterministic case, the argument presupposes 'a manipulability view of causation'.[47] Their idea is, as before, that if each effect represented in **V** can be manipulated separately from each other, CM2 will hold. But that is not the case. Think about the factory example. The level of the sewage-treating chemical can be set wherever we like just by introducing some cause other than the factory to produce it or to consume it; the same is true with the pollution level. If Hausman and Woodward can have external causes to do this job in the deterministic case, what stops me from having them in the indeterministic?

Consider the deterministic analogue to my chemical-factory example under the representation of intervention that Hausman and Woodward employ in their deterministic version of the argument – i.e. intervention by setting a special cause. We have two equations:

$$X \,c= C + U_X$$
$$Y \,c= C + U_Y$$

Hausman and Woodward tell us, recall, that given their other assumptions, U_X and U_Y each satisfy the requirements of an intervention, set-X and set-Y. So U_X and U_Y are being used to add to whatever C contributes to set X and to set Y at whatever values we may envision.

In the deterministic case C definitely contributes whatever value it takes. Let us imagine that it operates probabilistically instead, say 80 per cent of the time, both in producing X and in producing Y. We can represent this by altering the equations a little:

$$X \,c= \alpha_X C + U_X$$
$$Y \,c= \alpha_Y C + U_Y$$

where $Pr(\alpha_X = 1) = .8 = Pr(\alpha_Y = 1)$ and $Pr(\alpha_X = 0) = .2 = Pr(\alpha_Y = 0)$. So far I have made absolutely no commitment about CM2 one way or another. I have simply taken Hausman and Woodward's view about what the deterministic case must look like and modified it to allow that the common cause does not always contribute but acts, rather, probabilistically.

Our model is not very complete yet. We still need a joint probability for $U_X, U_Y, \alpha_X, \alpha_Y$. Call it P. Given that the α's are just there to represent the purely probabilistic behaviour of C, we can suppose that $P(\alpha_X \alpha_Y U_X U_Y) = P(\alpha_X \alpha_Y) P(U_X U_Y)$, where the joint probability for U_X and U_Y should factor, in accord with CM1. What can we say about $P(\alpha_X \alpha_Y)$? Well, we know it is perfectly easy to fix it so that CM2 is satisfied. Just insist on CMC: $P(\alpha_X = 1 \ \& \ \alpha_Y = 0) P(\alpha_X = 0 \ \& \ \alpha_Y = 1) = P(\alpha_X = 1 \ \& \ \alpha_Y = 1) P(\alpha_X = 0 \ \& \ \alpha_Y = 0)$. Suppose we do so. Then CM2 holds. Suppose next that we shift the joint probability just a little so that this equality does not hold. What is the big difference?

We had independent disruptibility in the deterministic model, in accord with Hausman and Woodward's demands. I keep it in the probabilistic model by assuming that whatever causes they had at work besides C are still at work – the only change is that C itself no longer contributes to the effect 100 per cent of the time. Now we begin to jigger P so that CMC holds, then it doesn't, then it does, but do not change anything else. CM2 will fluctuate in accord, but all the time each effect will be separately manipulable – by just exactly the same causes that Hausman and Woodward insist must be there to manipulate them in the deterministic case.

CM2 is false then and if it is false it cannot have been proved. So what is wrong with the proof Hausman and Woodward offer? The answer is, I think, one phrase near the bottom of p. 556. Before we get to that, let me summarize what they do.

In this case they are not so explicit about what the background assumptions are as they were for the deterministic case. In fact they need all the same ones that we used in the deterministic arguments of section 8.4, except of course for the assumption about determinism. I propose then that we keep (iii)–(vii) and replace (i)+(ii) by

(i)+(ii)' $\forall V_i$ in $\mathbf{V}$ $(V_i \ c^= \ f_i(\alpha(Par(1))_i Par(1)_i, \ldots, \alpha(Par(n))_i Par(n)_i) + a_i U_i)$, where $\alpha(Par(1))_i, \ldots, \alpha(Par(n))_i, a_i = 0, 1$ and $\{Par(1)_i, \ldots, Par(n)_i\}$ includes all and only parents of V_i.

The only other change they suggest is to replace MOD by a probabilistic version of it:

PM: $Pr(Y/Par(Y)) = Pr(Y/Par(Y) \ \& \ set\text{-}Z)$ where Z is any set of variables distinct from Y ....[48]

As with MOD they think PM is a necessary condition for a causal system; whereas I, as with MOD, think it is not only unnecessary but false – though it is an essential condition to add to the definition of 'intervention'. From PM they argue to MOD* in an unmodified form. My remarks about the argument are identical to those for the deterministic case. Next they try to argue from MOD*

[48] Hausman and Woodward (1999), p. 573.

to ($=$); and again my remarks are identical – except for one important addition that makes it impossible for the general argument to go through in even the ways sketched in the deterministic case.

Hausman and Woodward maintain, 'in the circumstances in which Par(X) is unchanging, either X varies spontaneously or because of causes that have no causal relation to any other variables except in virtue of causing X'.[49] I agree with this entirely, and it can readily be confirmed by inspection of the formula in assumption (i)+(ii)′: the variation in X conditioned on $Par(X)$ is due entirely to the combination of the spontaneous variation of X (represented by the α's) and the variation in U_X, conditioned on $Par(X)$.

Immediately next we are told, '[i]n both cases, given CM1, changes in X count as interventions with respect to [X]'.[50] Where does this assertion come from? It is not true, nor does it follow from anything previously assumed. It is true for U_X, and that it is true follows from the background assumptions. But it does not hold when 'X varies spontaneously', nor do our background assumptions imply that it should.

We cannot say that changes in X produced by 'spontaneous variation' will 'count as interventions with respect to X'[51] for we know that for an intervention in X, X should be produced by a method for which the resulting X values are probabilistically independent of any other quantities that are not effects of X. But just look at my factory example: unless CMC holds, the values of X produced by the common causes of X and Y will be probabilistically dependent on the values of Y thus produced. Where common causes are at work, the results of 'spontaneous variation' definitely do not behave like the results of intervention.

We must of course not be misled by the appearance of α in our probabilistic equations into thinking α can be used to set the value of X or Y, for after all α is just a piece of notation. It is a random variable, but that does not mean that it is used to represent some quantity. It is simply a notational device that we must introduce because of the odd way in which we have chosen to represent probabilistic causal laws. In line with our conventional methods for representing deterministic causality, we represent these laws as equations – equations between values of the effect and the values of the causes. But this does not reflect anything in nature. The whole point about probabilistic causality is that there is no equation between the values of the causes and that of the effect they produce. All the information that the equations convey is about the

---

[49] Hausman and Woodward (1999), p. 576.

[50] Hausman and Woodward (1999) p. 576. Note that they say 'with respect to Y'. But I think they must mean 'with respect to X' both because of the argument and because they later call these changes 'set-X'.

[51] I take it this means something like 'will have all the same probabilistic relations to other quantities as changes produced by an intervention'.

joint probability measure over the cause and effect quantities. This method of representation has some conveniences over simply writing down the measure, but it does not commit us to any strange new quantities.

8.7   'Cartwright's objection' defended

In the section titled 'Cartwright's objection', Hausman and Woodward argue,

[w]hen Y is produced by an intervention, Y carries no information about whether X will occur, but when Y is produced by C it does carry such information. Thus knowing that Y is produced by C provides information about X over and above any information that is contained in the full specification of C and of Y itself. It is hard to understand how this is possible.[52]

I disagree. It is not hard to understand – and indeed we understand in exactly the way Hausman and Woodward go on to suggest: 'If what is informationally relevant to X is the fact that Y has been caused by C, then it looks like it can only be a fact about some feature of (the causal structure or behavior of) C' (p. 569). What is informationally relevant about *Y* depends on *this* specific feature of the causal behaviour of *C*: *C* produces its effects in accord with the probability measure P (where $P(+X + Y/ + C) = .8$, $P(+X - Y/ + C) = P(-X + Y/ + C) = 0$, and $P(-X - Y/ + C) = .2$). Or, to point to what matters about this probability measure, *C* produces its effects in accord with a probability measure P that does not satisfy the very special restriction laid down in CMC.

The remainder of the interrupted sentence I just quoted from Hausman and Woodward goes on to say 'and hence a fact that one ought to take account of when one conditions on the common cause' (p. 569). This last does not follow. You cannot take account of how the common cause behaves just by conditioning on its occurrence.

Hausman and Woodward also maintain that

the only real-life cases that appear to have this sort of structure [the structure of the factory example] involve coarse macrovariables and are consistent with the satisfaction of the causal Markov condition at a more refined level of description (p. 568).

To this I have the same reply that I do to many proposals that insist that nature 'underneath' is different from how it appears to us. For many situations of interest our best models are probabilistic and many of these violate the causal Markov condition, including all those that posit side effects to the main effect. It is indeed consistent with these models that further refinement will allow us to produce new models that satisfy the causal Markov condition. Or models

---

[52] Hausman and Woodward (1999), p. 569.

that are completely deterministic. Or models that reduce all the phenomena of interest to physics. But that does not make these further metaphysical leaps a good bet. Of course we are equally not in a strong position to insist that the causal Markov condition is definitely violated, or that determinism or physics reductionism are not true.

What then should we do? To decide a given case, we need to be clear what the issues are, what hangs on the metaphysical choice we make and what are the costs of type I versus type II error, and – if anything other than the purity of our beliefs is at stake – hedge our bets.

8.8   Metaphysical defences of the causal Markov condition

Besides their arguments that MOD and strong independence imply the causal Markov condition, Hausman and Woodward also offer a number of separate metaphysical positions that would support it. Position (1) takes the operation of a cause to produce its effect as a real event that can be conditioned on to restore the causal Markov condition; (2) maintains that effects produced dependently are not really distinct effects; (3) uses the claim that probabilistic causation really involves the deterministic causation of chances and (4) is the thesis that underlies their use of the modularity condition, that each separate effect must be produced by a separate mechanism.

(1) Recall my discussion of the chemical factory, C, which operates purely probabilistically: the chemical, X, and the pollutant, Y, are always produced together, but they occur only 80 per cent of the time the production process is in place. Hausman and Woodward object to an earlier presentation of this example. '[Cartwright] speaks of C's "operation" or "firing" to produce X and Y. This description plays an important rhetorical role in making it seem intuitively plausible that there is no reason why the firings to produce chemicals X and Y must be uncorrelated when one controls for the common cause' (p. 562). They label a joint 'firing' F, and 'firings' to produce X and Y, F_1 and F_2 respectively. They go on to argue that there seems to be no sensible way to think of the relations of these F's to the other variables that preserves CM but violates the causal Markov condition.

I agree entirely. But this does not save the causal Markov condition because we should not think there are any such events as these F's. I used the expression 'C fires to produce X' for a specific purpose. If an effect X and its cause C are separated enough in space and time, they may well be linked by a chain of causal laws: $Cc \rightarrow C_1(X)$, $C_1(X)c \rightarrow C_2(X), \ldots, C_n(X)c \rightarrow X$; similarly for the effect Y. Each step in these two chains may itself be purely probabilistic. By talking about 'the firing' of C to produce X being correlated with 'the firing' of C to produce Y, I meant to ensure that I was only committing myself to the claim that $C_1(X)$ and $C_1(Y)$ are dependent given C, not for instance that $C_2(X)$ and $C_2(Y)$ are dependent given $C_1(X)$ and $C_1(Y)$.

My other use of 'firing' is in my informal account of why knowing that Y occurs can give us information about X. Hausman and Woodward are right to object if someone is moved to give up the causal Markov condition on the grounds that the occurrence of Y shows that a firing by C took place and that the firing guarantees the occurrence of X. Y gives us information about X just because of facts about the joint probability of X and Y conditional on C. Given C, X occurs if and only if Y occurs. Nor need the situation be so stark. For instance, Y might be only a probabilistic side effect of X (i.e. given C, $Pr(Y/X) < 1$). Still, Y is informationally relevant and for the same kind of reason – given C, there will be more X's when Y occurs than when it does not.

We should resist postulating events like the F's for all the well-known reasons, whether the causation involved is deterministic or purely probabilistic. Suppose C at t_0 causes E at t_1. Besides the events V_1: C occurring at t_0 and V_2: E occurring at t_1, is there an additional event, V_3: C's occurrence at t_0 causing E's occurrence at t_1? If so, what causes the causing? It looks as if we face an infinite regress unless we deny that all events have a cause.

There are also the usual worries about when the causing event occurs.[53] I will recall just one example. When does V_3 occur? If it occurs at t_0, the causing of E is entirely over and E has not yet occurred. What then causes it to occur? If V_3 occurs after t_0, the causing of E by C occurs when C no longer exists. If V_3 occurs throughout the period of both V_1 and V_2, still the causing of E by C goes on after C no longer exists as well as before E exists.

If we do feel compelled to admit all three of V_1, V_2 and V_3, I think the best solution is to follow Judy Jarvis Thompson[54] and take V_1 and V_2 to be parts of V_3, though that seems to reverse the empiricist's intuition. But this does not help Hausman and Woodward's case. CM1 obtains and without the need to postulate a cause for the causings. That is because the causings are not distinct events since they share as a part the occurrence of C. The causal Markov condition fails because the causings cannot serve as parents of the effects. Indeed, they cannot do so on any account on which they are not prior to the effects.

(2) In their section 8.9 Hausman and Woodward tell us of alpha particle emission, '[t]he event E consisting of the emission of two protons and the event F consisting of the emission of two neutrons are not related as cause and effect and are not independent conditional on the decay event'. This does not violate the causal Markov condition 'because there is really just a single effect ... and a single mechanism'. They add, '[a] similar point applies to macroscopic

---

[53] Cf. Russell (1913).   [54] Thompson (1977).

indeterministic phenomena which are produced by a single causal process, if there are any'.[55]

We can, I imagine, adopt some criterion for the individuation of event-types along these lines, though it will take some working out.[56] Then the causal Markov condition can be maintained for cases of probabilistic production of effects and side effects. What is important, of course, is that we not switch back and forth between this criterion of individuation and the usual one for which every separately measurable quantity defines an event type. We can have the causal Markov condition – but not for studying causal relations among the kinds of quantities we are generally interested in.

Besides the fact that this defence gives us a causal Markov condition that is of limited utility, it has a further methodological/epistemological drawback. The criterion of event individuation now involves facts about both what causes an event and how. This makes the job of causal inference enormously harder. Usually we start out with a set of event-types independently identified; then we look for the causal laws connecting them – which is hard enough. Under this proposal, we have to get the whole scheme all at once. This is a difficulty I take it that Hausman and Woodward acknowledge. In their conclusion they tell us that the causal Markov condition will be hard to apply: 'One needs a great deal of knowledge – indeed much more knowledge than may be available. It is often far from obvious how to divide some system into distinct causal mechanisms'.[57] They indicate the same thing earlier as well: 'One needs to know a great deal before one can justifiably assume that the Markov condition is satisfied ... One ... needs to know how to segregate the system correctly into distinct causal mechanisms' (p. 531).

(3) In their section 10 Hausman and Woodward maintain that 'if **CM** holds in deterministic circumstances, then, given a plausible assumption about what indeterministic causation consists in, it must hold in indeterministic circumstances as well' (p. 570). The plausible assumption is that 'X is a probabilistic cause of Y iff X is a deterministic cause of the chance of Y, ch(Y), where this is identified with the objective probability of Y' (p. 570).

For their argument Hausman and Woodward suppose that

$$(1) \ \neg(Xc \to Y).$$

[55] Hausman and Woodward (1999), p. 564.
[56] For instance, when E and F are produced dependently by A and F and G dependently by B, the single E/F event will overlap the F/G event, which may make problems for other claims about causal laws and procedures for inferring them. We will also need to be told how to treat the separate effects of E and of F. At the very least we will lose a great deal of information if we no longer represent E and F separately. Moreover, the proposal seems most plausible when the two effects are produced in total correlation, less so when the correlation is less than one.
[57] Hausman and Woodward (1999), p. 570.

They argue:

Then from the claim that probabilistic causation is deterministic causation of probabilities, one can infer that

(2) X is not a cause of ch(Y).

If CM holds for deterministic relations, it follows that

(3) $Pr(X/Par(X)\&ch(Y)) = Pr(X/Par(X))$.

Since (by **CM1**) Y will be independent of everything that ch(Y) is independent of, one can conclude for indeterministic variables X and Y, that if X does not cause Y, then $Pr(X/Par(X)\&Y) = Pr(X/Par(X))$.[58]

I have trouble understanding exactly what the argument here is. Still, I think it must be mistaken. To show this I will present a three-variable model that, I take it, satisfies their assumptions and for which their conclusion is false or else – under a less plausible interpretation of 'parents' – is true but will establish only a form of the causal Markov condition that is again of limited use.

I note first that Hausman and Woodward do not develop their notion that probabilistic causation of Y is equivalent to deterministic causation of $ch(Y)$. In particular it is hard to proceed in the discussion without answers to three questions. First, what will be the direct causes or parents of a quantity V represented in **V**? Will these be the probabilistic causes, i.e. the ordinary quantities we have been thinking about all along; or will ch(V) be the cause? Second, how does $Pr(\ldots ch(V) \ldots)$ relate to $Pr(\ldots V \ldots)$? Do we, for instance, have any principles like $Pr(V/ch(V) = r \& \ldots) = r$? Third, how does $ch(\ldots)$ relate to $Pr(\ldots)$ in general, and how do both relate to relative frequencies, either real or limiting? Happily for us, I do not need to make many assumptions about answers to the last two questions to establish my point.

Here is a model for the three-variable chemical-factory example that satisfies Hausman and Woodward's assumption that C is a probabilistic cause of E iff C is a deterministic cause of $ch(E)$. The probabilistic laws are as before, but I drop the U's for simplicity.[59]

Probabilistic causal laws

$$X \, c^= \alpha_X C$$

$$Y \, c^= \alpha_Y C$$

$$Pr(\alpha_X = 1 \& \alpha_Y = 1) = .8; \, Pr(\alpha_X = 0 \& \alpha_Y = 0) = .2$$

---

[58] Hausman and Woodward (1999), pp. 571–2.
[59] Recall, CM2 is true under Hausman and Woodward's background assumptions whether the U's are there or not. An example is easy to construct if one wants to include them.

Deterministic laws

$$ch(+x + y)c = .8\delta_C$$
$$ch(+x\neg y)c = 0$$
$$ch(\neg x + y)c = 0$$
$$ch(\neg x\neg y)c = .2\delta_C$$

where $\delta_C = 1$ when $+c$ and 0 when $\neg c$, and we assume that $ch(\ldots)$ satisfies the axioms of probability.

Now we ought to specify $Pr(X, Y, C, ch(X), ch(Y), ch(C), ch(XY), ch(YC), ch(XC), ch(X, Y, C))$. But my objection under the first interpretation of 'parents' will follow for any specification so long as we assume $Pr(\pm x \pm y/ \pm c)$ is equal to the value of $ch(\pm x \pm y)$ determined by $\pm c$ under the deterministic laws. For consideration of the second interpretation, I propose we suppose $Pr(X/ \ldots \& ch(X) = r) = r$, which is an assumption that favours CM2.

Let us now assume that the cause of X that is relevant for Hausman and Woodward's formula (3) is the probabilistic cause C. This is the nicer assumption since it looks as if the alternative choice of $ch(X)$ will generate a conundrum: $ch(X)$ could, presumably, cause X only probabilistically; hence it must deterministically cause $ch(X)$, which seems contrary to the normal assumption that nothing causes itself.

What I wish to show is that although Hausman and Woodward's penultimate step, (3), is true wherever the conditional probabilities are well defined, their conclusion, CM2, is false. Consider, for evaluating formula (3) in their argument, $Pr(X \& C \& ch(Y)) = Pr(X \& C)$ if the value of $ch(Y)$ is that determined by C, and 0 otherwise. Similarly $Pr(C \& ch(Y)) = Pr(C)$ if the value of $ch(Y)$ is that dictated by C, and 0 otherwise. So $Pr(X/C \& ch(Y)) = Pr(X/C)$ wherever the conditional probabilities are defined.

Now consider $Pr(+x \& +y \& +c)/Pr(+y \& +c) = Pr(+x \& +y/ +c)/Pr(+y/+c)$. By my assumption about the relations between chances and probabilities over the 'traditional' quantities, this gives $.8/.8 = 1$. But $Pr(+x/ +c) = .8$ since $ch(+x) = .8\delta_C + 0 = .8$ for $+c$. So CM2 is violated, despite the fact that in the model C probabilistically causes E iff C deterministically causes $ch(E)$. We should note that CM1 is satisfied: $\neg(X \perp Y)$ but X and Y have a common cause C and $ch(X) \perp ch(Y)$.

Finally, when the causes of X and Y are taken to be $ch(X)$ and $ch(Y)$, CM2 follows immediately given my assumption about the relation between $Pr(\ldots)$ and $ch(\ldots)$: $Pr(V/ch(V) = r \& \ldots) = r = Pr(V/ch(V))$. So here after all is a way to defend CM2. And that is fine – so long as we are careful to keep track of the fact that we are now taking chances to be causes. That means that, for CM2 to be of any practical use, we will have to figure out not only how to measure the causes – what is the value of $ch(V)$? – but also how to find out probabilities like $Pr(ch(X)ch(Y)ch(X\&Y)XY)$ – which are hard even to interpret, let alone measure.

(4) Hausman and Woodward place three superficially distinct requirements on a system of causal laws, all of which get treated as more or less the same in their discussion. The first is the claim that it must be possible to manipulate each cause separately. The second is the claim, which plays an important role in Hausman's earlier work, that each effect E has a cause that has no cross-restraints with any causes of any other effect, except for factors that E causes. The third, which is a focus of other work by Woodward, is that each effect must be produced by a separate mechanism. The meeting ground for Hausman and Woodward seems to be their feeling that the two conditions they focus on separately are in fact equivalent to the first, the requirement of independent manipulability.

This requirement of independent manipulability is the claim that Hausman and Woodward offer as the underlying motive for their modularity condition: the section introducing modularity begins, '[o]ne crucial fact about causation ... is that causes are as it were levers that can be used to manipulate their effects'.[60] But it is the second requirement that appears in their proof that modularity implies the causal Markov condition. The step from one to the other is odd. For the first demands that every cause be manipulable (and, as Hausman and Woodward further develop it, separately manipulable), whereas the second requires that every effect be separately manipulable.

Consider my simple three-variable example of the chemical factory. We may assume that the first claim is readily satisfied – if we increase the number of hours the factory works each day, or decrease it, we thereby manipulate how much chemical results; and we can in the same way manipulate how much pollutant results. Whether we can do that or not is independent of whether the pollutant and the chemical are manipulable separately from each other. As we have seen in section 8.4, for the interpretation of 'intervention' that Hausman and Woodward use in their proof, the separate manipulability of the chemical and the pollutant depends on whether each has the right kind of unrepresented cause, and the manipulability of C is not relevant to that.

What about the other kind of intervention, though, the kind where a factor is fixed not by manipulating a special cause to set it where we will, but rather by changing the law for its production? Here at least I can begin to share Hausman and Woodward's intuition. They say that, barring other causes of the chemical and the pollutant, the two are produced by the same mechanism. They suggest that it follows that the two are not separately manipulable, and thus violate a

---

[60] Hausman and Woodward (1999), p. 533.

more general version of the second condition described just above: every effect should be separately manipulable.

I say that I begin to share their intuition, but I have objections at all three stages. My objections to the last step are strongest. I do not see why every effect should be separately manipulable. Indeed, as we have seen, I do not even agree that every cause should be separately manipulable. But how do we get from the demand that causes be separately manipulable to the demand that effects be so as well? Hausman has independent arguments that they will be, which I have discussed in ch. 7. Here I am concerned with the step from the first to the second condition.

One way that the connection can be drawn is via methodology. If we can intervene in the level of the chemical and thereby alter the level of pollution, we would have a good argument that the chemical causes the pollution. But in this case we are not using the metaphysical demand that causes are levers for their effects, but rather a demand that nature be nice to us epistemologically: every epistemically possible cause should be separately manipulable, so that we have a nice method for finding out if it is a cause of the effects of concern. The one is no guarantee of the other.

Now turn to the second step. If we think that C produces X and Y by the same mechanism, does it follow that X and Y are not separately manipulable? Certainly not if we countenance manipulation either by causes other than C, as in section 8.4, or by interventions into the causal processes linking C with X or C with Y. But what if we are solely concerned with what in section 8.2 I called 'causal-law intervention'? If C produces X and Y by the same mechanism, can the law for the production of X be changed without changing that for Y? I take it this question means, 'can the probabilistic causal laws for X and Y recorded in section 8.8 in the discussion of argument (3) above be changed separately from one another?' I do not know the answer, nor even how to set about finding it.

Here is one way to generate a mistaken answer. In the example under discussion, the two effects are produced in total correlation. So we may suppose α_X is identical to α_Y. In this case, it is logically impossible to change α_X without changing α_Y. This argument is mistaken on several counts. First, we know that we should not be reifying α_X and α_Y. Nor should we be reifying 'the mechanism' as some feature over and above the cause, its effects, and the causal laws linking them.

But we can recast the question without this apparent reification. $\alpha_X = \alpha_Y$ is simply a way of recording a fact about the joint conditional probability, $Pr(XY/C)$. We know that in the given system of causal laws, this joint probability violates CMC. It seems we wish it to obtain in the new system. Can it?[61]

---

[61] Throughout this discussion we are always talking about setting the level of quantities and not, for instance, their probability. So we want in the new system $Pr(+x) = 1$ or $Pr(-x) = 1$. This trivially implies CMC. Similar considerations to those I raise in the text can also be made for the case where we set the probabilities and not the levels.

I still do not know how to answer the question. And I suspect it has no general answer. How we can affect the probabilities with which effects are produced by a given cause governed by a given system of laws will depend very much on what the cause is and how this particular system of causal laws arises.

Last, I should like to recall to mind the vast number of cases where the effect and the side effects are not produced in perfect correlation but are still probabilistically dependent on one another given the cause – for instance, where the side effect only occurs some fixed percentage of the time the effect occurs. Do we want to say the two are produced 'by the same mechanism'?

Of course what we should say will depend heavily on what consequences we intend to draw from a verdict either way. But in so far as I feel Hausman and Woodward's pull to say it is the same mechanism when the correlation is total, it seems equally right to say that it is the same mechanism when the correlation is only partial. But then I think the same thing holds when there is no correlation and CMC obtains. For the two are still produced by the same cause operating under laws that set the joint probability for the entire outcome space. On the other hand, 'same mechanism' could simply mean 'CMC obtains', but in that case we do not have an independently based intuition from which to defend the causal Markov condition.

8.9   Conclusion

Is the causal Markov condition true of all probabilistic causal systems? If I am right in section 8.4, modularity is after all no argument for it. Nor, I think, are Hausman and Woodward's other defences strong enough to secure it. The best strategy then, I would urge, is to assume it only for the systems where we have reasons specific to the system itself to suppose it holds.[62]

APPENDIX   PROOF OF CM1 FOR LINEAR DETERMINISTIC SYSTEMS WHERE THE EXOGENOUS VARIABLES ARE INDEPENDENT IN ALL COMBINATIONS

Claim 1
Suppose

$$X_1 \, c= U_1$$
$$X_2 \, c= A_{21} X_1 + U_2$$
$$\vdots$$
$$X_k \, c= \sum_{i=1}^{k-1} A_{ki} X_i + U_k$$

---

[62] This discussion is continued in Hausman and Woodward (2004) and ch. 9 of this book.

We can re-express this as follows:

$$X_k \, \mathtt{c} = \sum_{i=1}^{k} U_i \sum_{l=i}^{k-1} A_{kl} \sum_{m=i}^{l-1} A_{lm} \sum_{n=i}^{m-1} A_{mn} \cdots$$

where we adopt the convention, $\sum_{j=\alpha}^{\beta} f_i(j, k, l, m \ldots) = 1$, if $\alpha > \beta$. Thus we arrive at the reduced form $\alpha > \beta$:

$$\mathrm{RF}: X_k \, \mathtt{c} = \sum_{i=1}^{k} \Gamma_i^k U_i$$

where

$$\Gamma_i^k = \sum_{l=i}^{k-1} A_{kl} \sum_{m=i}^{l-1} A_{lm} \sum_{n=i}^{m-1} A_{mn} \cdots$$

Claim 2
Given

all combinations of U_i's are independent of all others   (1)

$$X_k = \sum_{i=1}^{k} \Gamma_i^k U_i \tag{2}$$

$$X_h = \sum_{j=1}^{h} \Gamma_j^h U_j \tag{3}$$

Γ defined as above

then $P(X_k = x \wedge X_h = y) = P(X_k = x) \cdot P(X_h = y)$ iff for some i, $\Gamma_i^k \cdot \Gamma_i^h \neq 0$

Proof

$P(X_k = x \wedge X_h = y)$

$\quad = P\left\lfloor \{\vee\{\wedge_{\{u_i\}\ni X_k=x}(U_i = u_i)\}\} \wedge \{\vee\{\wedge_{\{u_j\}\ni X_h=y}(U_j = u_j)\}\} \right\rfloor$

$\quad = P\left\lfloor \vee\{\{\wedge_{\{u_i\}\ni X_k=x}(U_i = u_i)\} \wedge \{\wedge_{\{u_j\}\ni X_h=y}(U_j = u_j)\}\} \right\rfloor$

$\quad = \sum P(\wedge_{\{u_i\}\ni X_k=x}(U_i = u_i)) \cdot P(\wedge_{\{u_j\}\ni X_h=y}(U_j = u_j))$

iff for some i, $\Gamma_i^k \neq 0$ *and* $\Gamma_i^h \neq 0$ (i.e. the same U_i appears in the expansion for both X_k and X_h)

$\quad = \left(\sum P\{\wedge_{\{u_i\}\ni X_k=x}(U_i = u_i)\} \right) \cdot \left(\sum P\{\wedge_{\{u_j\}\ni X_h=y}(U_j = u_j)\} \right)$

$\quad = P(\vee \wedge_{\{u_i\}\ni X_k=x}(U_i = u_i)) \cdot P(\vee \wedge_{\{u_j\}\ni X_h=y}(U_j = u_j))$

$\quad = P(X_k = x) \cdot P(X_h = y)$

Claim 3

The following now follows trivially:

in a linear deterministic system in a block triangular array with independence of exogeneous variables in all combinations,
CM1: X_k and X_h are not independent of each other →

$$X_k \text{ c} \rightarrow X_h \text{ or } X_h \text{ c} \rightarrow X_k \text{ or } \exists X_d \ni X_d \text{ c} \rightarrow X_h \text{ and } X_d \text{ c} \rightarrow X_k$$

For they will be independent unless some U_i in the equation (2) for X_k is the same as a U_i in the equation (3) for X_h; i.e. unless either they share a common parent, X_d, or one of them causes the other.

9   From metaphysics to method: comments on manipulability and the causal Markov condition

9.1   Introduction

Metaphysics and methodology should go hand in hand. Metaphysics tells us what something is; methodology, how to find out about it. Our methods must be justified by showing that they are indeed a good way to find out about the thing under study, given what it is. Conversely, if our metaphysical account does not tie in with our best methods for finding out, we should be suspicious of our metaphysics.

Daniel Hausman and James Woodward try to forge just such a connection in their work on causation. They claim that the central characterizing feature of causation has to do with manipulability and invariance under intervention. They then use this to defend the causal Markov condition (CMC), which is a key assumption in the powerful Bayes-nets methods for causal inference. In their own words, 'the view that causes can in principle be used to control their effects lends support to the causal Markov condition'.[1] This is an important project and, to my mind, a model of the kind of thing we should be trying to do. Their first attempt to prove a link between manipulability and CMC[2] had a number of problems however.[3] Unfortunately, so too does their latest attempt, 'Modularity and the Causal Markov Condition' (hereafter M&CMC).[4]

Although the connection they picture is just the kind we need between metaphysics and method, this particular link is just not there. The first reason is that the premise they start from has nothing to do with the fact that causes can be used to control their effects. Instead it, at best,[5] lays out a sufficient condition for inferring a causal relation in ideal experimental tests; and taking such a condition as part of the metaphysics of causality, as central to the very idea of causality, smacks too much of operationalism. The most blatant of the problems with their project in M&CMC, however, is that the proof is not valid, at least

Research for this paper was carried out in conjunction with the AHRB project, *Causality: Metaphysics and Methods*, and supported by grants from the Latsis Foundation, University of California at San Diego and the National Science Foundation, for which I am extremely grateful.

[1] Hausman and Woodward (2004), p. 148.   [2] Hausman and Woodward (1999).
[3] Cartwright (2002).   [4] Hausman and Woodward (2004).
[5] See discussion in section 9.6 for why I say 'at best'.

under what seems to me the most natural reading of it. On a second reading, the premise is blatantly false and on a third, the proof is again invalid.

I shall explain the problems with the proof after a review of the switch they have made in their work from taking control, or manipulability, to taking a sufficient condition for inferring a causal relation in an experimental test as their starting point. The final discussion will focus on cases of probabilistic causality. When causes can act probabilistically, CMC will be violated in any case where causes produce by-products in tandem with their main effects. Hausman and Woodward maintain that causes cannot do that. I shall defend my view that there is nothing to stop them from doing so.

First a definition and some notation. The causal Markov condition is formulated relative to a population Φ, a set of random variables $\mathcal{V}$ on that population, a set of random variables $\mathcal{U}$ representing omitted causes of features represented in $\mathcal{V}$ sufficient in combination with the variables in $\mathcal{V}$ to fix the values (or, for indeterministic cases, the chances) of every variable in $\mathcal{V}$, a probability measure P over $\mathcal{V} + \mathcal{U}$, and a directed graph G of the causal relations among the features represented by variables in $\mathcal{V} + \mathcal{U}$:[6]

CMC: $\Phi, \mathcal{V}, \mathcal{U}, P, G$ satisfy CMC iff for all X_i, X_j, $i \neq j$, in $\mathcal{V}$, if X_i does not cause X_j, then X_i and X_j are probabilistically independent conditional on pa_i (i.e. $X_i \perp X_j / pa_i$),

where pa_i is the set of direct causes, or parents, of X_i relative to $\mathcal{V}$ and G.

As to notation, throughout I shall use $X c \to Y$ to mean that X causes Y and $X c= f(\ldots)$ to indicate that the factors in the function on the right-hand side cause those on the left and that the functional equality holds, where in both cases generic-level as opposed to singular causation is intended. I shall denote members of $\mathcal{V}$ by X_i or Y_i, values of variables by lower case versions of the letter representing the variable, and a member of $\mathcal{U}$ that causes X_i by U_{ij}. Following Hausman and Woodward, I shall use U_i' to represent the net causal effect on X_i of a minimal set of omitted causes of X_i that in combination with the parents of X_i are sufficient to fix the value (or for indeterministic cases, the chance) of X_i. $A \perp B/C$ means that A is probabilistically independent of B conditional on C.

Hausman and Woodward treat CMC for purely deterministic causality and for purely probabilistic causality in one fell swoop. I shall divide my discussion to focus on different aspects of the proof.

---

[6] Usually it is said that the graph is over $\mathcal{V}$, but often in practice U's appear on the graphs as exogenous causes. This is particularly important for Hausman and Woodward since in their proof the interventions on variables in $\mathcal{V}$ – which are supposed to be causes of the quantities represented by those variables – will be members of $\mathcal{U}$. There is also the question of whether various concepts like intervention are defined relative to a graph or to 'reality'. Since concepts central to CMC are defined relative to a graph, I think it is best to define all the concepts relative to a graph. (The alternative is at best very messy and certainly impossible without resorting to the concept of a 'correct' graph.)

9.2   Earlier views: manipulability versus testability

Hausman and Woodward have long defended the idea that modularity is a characterizing feature of causality and this term appears in the title of the paper with the new proof – 'Modularity and the Causal Markov Condition'. In much of their earlier work modularity was intimately connected with claims that at the heart of causation is the idea that something is a cause just when it can be used to control the putative effect. I want here to review the earlier ideas to make clear that their new proof does not link CMC with modularity as we first saw them talking about it. Nor do they claim so when they write down their central premise in M&CMC. Still, it is easy to be misled since they retain the earlier language as well as a number of the earlier slogans, such as the one quoted in section 9.1 above claiming that the fact that causes can be used to control their effects supports CMC.

I shall not start by defining modularity because I think some of the arguments in their earlier work, including the paper where the earlier proof appears, speak to a somewhat different thesis than the one they formally state as MOD in the earlier proof.[7] Rather I shall describe two motivations for modularity we can find in their work, motivations that lead to different conditions.

9.2.1   Manipulability

It is essential to causality[8] that causes can be used to manipulate their effects.[9] So (roughly) . . .

$(Xc \rightarrow Y) \rightarrow$ there is some way (they call it an 'intervention' or sometimes a 'manipulation') to change X so that Y changes.

Both 'intervention' and 'manipulation' suggest human agency and indeed for many philosophers that has been important. This, however, is not part of

---

[7] Cf. my discussion of MOD versus MOD# (see ch. 8 of this book).

[8] Hausman claims that his work is intended to provide a boundary condition for the applicability of causal concepts. In this case my remark here should read, 'It is essential to the correct application of "cause" that . . .'. Perhaps thereafter we need always to read '$Xc \rightarrow Y$' as 'It is correct to say that X causes Y'.

[9] Cf. Woodward's principal claims/definitions in Woodward (2003), TC, DC and M. All state as necessary and sufficient condition for X to be some particular kind of cause of Y, that 'there is a possible intervention on X that will change Y'. Again (in 2003, p. 114), Woodward explains that these conditions 'tell us what must be true of the relationship between X and Y if X causes Y'. It is in this very same paragraph, however, that we see a withdrawal from the strong claims recorded in TC, DC and M, that causality requires that there *be* interventions:

When we engage in causal inference regarding the effects of X in a situation in which there is no variable that satisfies all the conditions for an intervention variable with respect to X . . . we should think of ourselves as trying to determine what would happen in an ideal hypothetical experiment in which X is manipulated in such a way that the conditions in [the definition of intervention] *are* satisfied. (2003, p. 114, italics original)

Hausman and Woodward's programme. What they require is not that a cause be manipulable by us in the right ways but merely that it be possible that the cause vary in the right ways, whether we vary it or not. This is a theme familiar from the literature on natural experiments, i.e. situations in which one factor varies naturally, without our help, in just the right way to count as a test for causality. Hausman and Woodward are explicit that human agency is not required. Nevertheless 'manipulation' and 'intervention' are the words they regularly use rather than a more neutral description in terms of variation. So we must be careful to focus on the definitions themselves and not the labels.

Even with this understanding of what 'manipulability' means, the condition seems far too strong. If a cause can vary in the right way, then (for the most part[10]) we can expect its effects to change in train. But there is no guarantee that such variation is always possible.[11]

9.2.2   Testability

In discussing the chemical factory example described below in section 9.9, Hausman and Woodward take it to be an advantage of their view that it allows one

to disentangle different possibilities concerning the causal structure of the situation. If one thinks of the example as one in which the effects cannot, even in principle, be separately interfered with, the example does not really have a common cause structure, but is rather one in which a single mechanism links [the putative common cause] to X [one putative effect] and Y [the other putative effect]. . .[12]

Thus, it seems, the case would not have a common cause structure if there were no interventions possible on X and Y, at least in principle, that would allow us to determine that it does, and to do so by showing that X does not cause Y or the reverse. This suggests that they want to require that for every possible causal connection, $Xc \rightarrow Y$, there should be (at least in principle) an intervention on X that would show whether it holds or not.

This suggestion is supported by the kinds of argument about causal mechanisms that they repeatedly offer in defence of modularity. Each causal principle is to represent a separate mechanism for the production of some given effect X. But there is no separate mechanism for X unless it is possible to intervene on X without changing any other causal principles. So again, it looks as if for every possible effect X there must be a possible intervention, and presumably this intervention should leave $P(Y)$ unchanged if X does not cause Y since the intervention is supposed to have no effect on any other mechanisms, either to add, subtract or change them. Again this demand is tantamount to a condition

---

[10] See caveat in next section.   [11] For a more extended discussion see ch. 7 of this book.
[12] Hausman and Woodward (2004), p. 159.

that each possible causal claim be testable, and testable by what I shall call 'experiment': intervene and see what happens.

So, the requirement of testability by experiment provides a second distinct way to formulate a modularity requirement. Using the notation and formulations of M&CMC, we have:

For all Φ, $\mathcal{V}$, $\mathcal{U}$, P, G, and X_i in $\mathcal{V}$, there is a Z_i in $\mathcal{V} + \mathcal{U}$ such that Z_i is an intervention on X_i and $\forall X_j \neg (X_i c \rightarrow X_j)$ iff $P(X_j / Z_i = \text{on}) = P(X_j / Z_i = \text{off})$.[13]

This requires explanation.

Z_i in $\mathcal{V} + \mathcal{U}$ is an intervention on X_i relative to Φ, $\mathcal{V}$, $\mathcal{U}$, G iff

 (i) Z_i causes X_i on G.
 (ii) Z_i is not caused on G by any of the other variables in $\mathcal{V} + \mathcal{U}$.
 (iii) Z_i does not on G cause any members of $\mathcal{U}$ and has no causes in common with any members of $\mathcal{U}$ or other Z's on G.
 (iv) For all $X_j, j \neq i$, if Z_i or any cause of Z_i causes X_j on G, then it does so only via a path passing through Z_i and X_i first.
 (v) If X_i is deterministically caused on G, P, then for some range of values of Z_i, $\mathbf{z}_i^*$, if $Z_i = z_i^*$ in $\mathbf{z}_i^*$, then $X_i = x_i^*$ regardless of the values of any other members of $\mathcal{V} + \mathcal{U}$. If X_i is indeterministically caused, then for some range of values of Z_i, $\mathbf{z}_i^*$, if $Z_i = z_i^*$ in $\mathbf{z}_i^*$, $P(X_i) = P_i^*$ regardless of the values of any other members of $\mathcal{V} + \mathcal{U}$. For other values of Z_i, X_i or $P(X_i)$ is a function of pa_i and members of $\mathcal{U}$.

The values in $\mathbf{z}_i^*$ are designated as the on values for Z_i. So the condition says that for every variable X_i there always is an intervention and that the probability of any other variable X_j changes when X_i is intervened on iff X_i causes X_j.

What is important to notice is that testability is stronger than manipulability in two ways:

• Testability requires that there exists an 'intervener'/'manipulation' for every factor, not just for causes.
• Under testability, manipulating a cause changes its effects; but also manipulating non-causes of a factor does not change it.

I think that Hausman and Woodward's views on causal mechanisms and interventions commit them to something like testability by experiment. As an attempt to motivate modularity, testability also has the advantage over manipulability that it has the arrows of implication going the right way. Manipulability says that if $Xc \rightarrow Y$ then intervening on X changes Y. The assumption they call MOD*, which is the premise in their argument for CMC, instead requires that

---

[13] Hausman and Woodward do not use the conditional probability, presumably because they do not wish to assume that the interventions themselves have a probability measure. But they need a measure over interventions to discuss MOD* since interventions are supposed to be probabilistically independent of various variables in $\mathcal{V}$, so I assume throughout that there is a measure over $\mathcal{V} + \mathcal{U}$.

if $\neg(Xc \rightarrow Y)$ then intervening on X does not change Y, which does follow from testability by experiment. To the extent that I am right, this is extremely restrictive: it not only requires that causal relations, in order to be causal, must each be ascertainable by us, but moreover that they be ascertainable by one specific method among the many that we use (like various 'mark methods'). This is operationalism pushed beyond its limits.

   These two considerations lead me to:

Conclusion 1: modularity in the form of either the manipulation or the testability thesis is too strong a condition to characterize causality.

9.3   Increasingly weaker theses

For the latest proof Hausman and Woodward do not start from testability but from a far weaker assumption about the metaphysics of causality. Why? Testability tells us that for every V and every X_i in V there is an intervention Z_i for X_i and Z_i changes X_j iff $X_i c \rightarrow X_j$.[14] This is too strong for at least two reasons:
       There isn't always such a Z_i. (This is my explanation; they don't
          themselves say this.)
       The 'iff' is too strong. If X_i both causes and prevents X_j then X_j
          need not change as X_i does. (They do say this, though, as I argue
          in section 9.4, I do not think they need to.)
What they propose instead in M&CMC is this: MOD* 'says that when X_i does not cause X_j, then the probability distribution of X_j is unchanged when there is an intervention with respect to X_i'.[15] So, roughly (for a more precise statement, see section 9.5)

if Z_i is an intervention for X_i then $(X_j$ or $P(X_j)$ changes under $Z_i) \rightarrow (X_i c \rightarrow X_j)$

   We should note that this gives up on:
       the claim that the possibility of full testability is necessary for the
          applicability of causal concepts;
       the claim that it must be possible to use a cause to manipulate its effects
   and it does so in two ways:
       • it is no longer necessary that an intervention on X_i exist in order for
         X_i to cause some other factor;
       • it does not require that manipulating a cause changes the effect but
         rather that if X_i does not cause X_j then manipulating X_i will not
         change X_j.

---

[14] They still maintain this thesis in places in M&CMC and still sometimes conflate it with the
   weaker MOD*. See note 24 below.
[15] Hausman and Woodward (2004), p. 149.

MOD* (or rather some more precise version of it as I discuss below) is the premise in Hausman and Woodward's new proof of CMC. So it seems they do not link manipulability with the causal Markov condition, but at best only a claim about one test that can guarantee that a causal relation holds. Nor do they deny this: in M&CMC they conclude 'The causal Markov condition is a probabilistic doppelganger for *invariance to intervention*'.[16] Still, they call the section with the proof, 'Causation and *manipulation*' (my italics) and begin it with:

> When X causes Y and one can intervene to change the value of X, one can use one's knowledge of the causal relation to influence the value of Y . . . This is an extremely important feature of causation. One way to formulate a connection between causation and manipulability . . . is to say that if an intervention with respect to X_i changes the probability distribution of some other variable X_j, then X_i causes X_j.[17]

Given that their central premise is MOD*, their proof may connect something about manipulability with causation but:

> The proof in M&CMC does not connect the claim 'Causes *can always* be manipulated to affect their effects' with causation,
>
> or the weaker claim 'If a cause can be manipulated (in the right way), the effect is changed'.

That is because MOD* says that if we manipulate a factor that is not a cause of another, the other does not change.

From this consideration and others in this section and the last I draw

Conclusion 2: the premise (MOD*) in their proof is not manipulability but at best one test that, *if* it can be applied and *if* it is passed, can guarantee that one factor causes another.

9.4   The proof is invalid

There are two points we need to beware of:

- The earlier Hausman and Woodward proof used the strong premise that for every X_i in $\mathcal{V}$ there is an intervention Z_i and manipulating Z_i leaves X_j unaffected if $\neg (X_i c \rightarrow X_j)$. The earlier proof did not work. The new proof has a weaker premise. How can it work?
- One would think that whether the probability of a non-effect of X_i, X_j, is left the same under an intervention on X_i will depend on whether the intervention on X_i is probabilistically dependent on any causes of X_j. Such dependencies are often prohibited by definition of 'intervention'. But not so for Hausman and Woodward. How can they get by without this?

[16] Hausman and Woodward (2004), p. 153, my italics.
[17] Hausman and Woodward (2004), pp. 148–9.

Let me recall a well-known result:

For any $\mathcal{V}$, (U_i' are independent of each other in all combinations $\rightarrow$ CMC).

So when are the U''s independent? Here is one common hypothesis:

CM1: factors that are not causally connected are independent in all combinations. (X is causally connected with Y iff $Xc \rightarrow Y$ or the reverse or they have a common cause.)

If $\mathcal{V}$ is causally sufficient, the U_{ij}'s will be causally unconnected and hence given CM1 independent in all combinations. This rules out 'brute correlations' that have no causal explanation, like
1 Elliott Sober's case of bread prices in Britain and sea levels in Venice.[18]
2 Any case with time trends.

In their new proof Hausman and Woodward do not assume CM1. So how do they rule out probabilistic dependencies that are incompatible with CMC? They think they can do so using MOD* plus two other assumptions, where, more precisely stated,

Φ, $\mathcal{V}$, $\mathcal{U}$, P, G satisfy MOD* iff for every X_i in $\mathcal{V}$ and every intervention Z_i in $\mathcal{V}+\mathcal{U}$ on X_i, $Z_i \perp X_j$ for any X_j such that $\neg(X_i c \rightarrow X_j)$.

Notice that as I have written it, MOD* is a condition that a system might satisfy, not a claim. So too is CMC. I shall be concerned about what claims Hausman and Woodward want to assert. One claim that many favour is that any representative causally sufficient system, Φ, $\mathcal{V}$, $\mathcal{U}$, P, G, satisfies CMC. Hausman and Woodward say, 'We shall show that MOD* . . . [and some other assumptions] . . . imply CMC'.[19] The most natural reading of this is that any system that satisfies MOD* plus the other assumptions satisfies CMC, and this is what I shall suppose they mean.

The two other assumptions for the case of determinism are

1 '[A]ll the variables in $\mathcal{V}$ are distinct, . . . we are dealing with the right variables, and . . . selection bias and other sources of unrepresentativeness . . . are absent'.[20]
2 $\mathcal{V}$ is causally sufficient.

So,

H-W-claim₁: for any Φ, G, $\mathcal{V}$, $\mathcal{U}$, P, *if A, B* and MOD* hold for Φ, G, $\mathcal{V}$, $\mathcal{U}$, P, then CMC holds for Φ, G, $\mathcal{V}$, $\mathcal{U}$, P.

Here is how their proof proceeds:
• Define 'intervention' so that interventions on X_i are causally unconnected with U_j' if X_i does not cause X_j.

---

[18] Sober (2001).   [19] Hausman and Woodward (2004), p. 149.
[20] Hausman and Woodward (2004), p. 148.

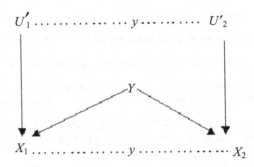

Figure 9.1

- Show that in a certain subpopulation – the subpopulation where pa_i is fixed – U_i' satisfies the definition of an intervention.
- Use MOD* to claim that in this population $U_i' \perp X_j$; i.e. $U_i' \perp X_j/pa_i$.
- It follows, they say, that $X_i \perp X_j/pa_i$.

But the proof must be invalid since there are cases that satisfy the premises but where CMC fails. Consider fig. 9.1 for some population Φ that satisfies A, B and MOD* and for which $X_i \, c^= a_i Y + U_i'$ ($a_i \neq 0$) and for which U_1 and U_2 are dependent conditional on Y. This system is inconsistent with CMC. (A dotted line indicates probabilistic dependence; a dotted line with a y through it, dependence conditional on Y; a dotted line with a slash through it, independence. In this example it makes no difference whether $U_1' \perp U_2'$.)[21]

In fig. 9.1 MOD* is satisfied: there is no factor in $\mathcal{V} + \mathcal{U}$ that sets the value of X_i, hence no intervention, so MOD* holds vacuously.

It is worth rehearsing just why the proof is invalid. The theorem we wish to prove says that if a given population Φ and a given probability P[22] satisfy MOD* then that population and that probability satisfy CMC for X_i, X_j such that $(X_i c \rightarrow X_j)$. Hausman and Woodward's argument establishes that if MOD* holds for some different population Φ' and different probability measure P', then CMC holds for Φ. (The population Φ' is the subpopulation in which pa_i is fixed and $P'(\ldots) = P(\ldots|pa_i)$.)

We can thus draw

Conclusion 4: the proof is invalid under the most natural reading of Hausman and Woodward's claim.

---

[21] In many treatments the situation pictured in fig. 9.1 is ruled out by CM1. But recall that Hausman and Woodward do not assume CM1.

[22] For brevity I here repress the other quantifiers and the other assumptions.

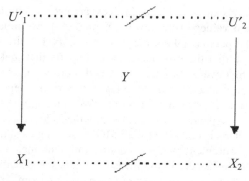

Figure 9.2

9.5   MOD* is implausible

We can also use fig. 9.1 to illustrate how strange the condition MOD* is, independent of its connection with CMC. Compare fig. 9.1 and fig. 9.2.

For fig. 9.2 suppose, as with fig. 9.1, that $X_i c = a_i Y_i + U_i'$ and A and B are satisfied. But for fig. 9.2 imagine that $a_i = 0$ and that $U_1' \perp U_2'$ as MOD* requires. So, MOD* allows U_1 and U_2 to be probabilistically dependent in fig. 9.1 but prohibits it in fig. 9.2.

That seems to require a completely *ad hoc* distinction between the two cases. Suppose we start with a situation appropriately represented by fig. 9.1, with U_1' and U_2' probabilistically dependent. Consider a situation identical with this except that Y's influence on X_1 and X_2 is just slightly less (i.e. a_1 and a_2 are slightly smaller). MOD* does not prohibit this new situation either. Now consider a series of situations in each one of which Y's influence on the X's is smaller than in the one before. Still MOD* does not prohibit the U''s from being dependent. This is true no matter how small Y's influence on the X's becomes, so long as it has any influence at all. But as soon as the influence disappears altogether ($a_1 = 0 = a_2$), suddenly under MOD* the U''s must be independent. What is responsible for this sudden jump?

We may even suppose that the diminutions of Y's influence occur across time in the very same physical system. Gradually Y's powers to influence X_1 and X_2 give out. What would ensure, when Y's influence finally disappears altogether, that suddenly U_1' becomes independent of U_2'? I see nothing that could.

Here is an example (or rather, a caricature of an example). Suppose Elliott Sober is correct that bread prices in England are probabilistically dependent on Venetian sea levels. We can suppose that the real levels of these two variables in combination with the measurement apparatuses employed (call this combination U_1 for sea levels and U_2 for bread prices) are each a central cause of the respective measured values of the levels (X_1 and X_2); presumably so too will be the skill of the persons taking the measurements. For the sake of an example let us suppose that there is one team of experts that make both such measurements and that every ten years more and more automated technology is introduced in both places so that gradually the results depend less and less on the skills of the measurement team (Y). We can suppose that U_1 and U_2 are probabilistically dependent because by hypothesis bread prices and sea levels are dependent. This is consistent with MOD* so long as skills matter. But as soon as the measurement process becomes fully automated and the skills of the team have no influence on the measured values, suddenly bread prices and sea levels, which were dependent until then, must become independent if MOD* is to be satisfied. I do not see why this kind of thing should happen.

Of course if we assume CM1, bread prices and sea levels will not be dependent in the first place. But recall that Hausman and Woodward do not assume CM1. And that is all to the good given their overall programme because, given CM1 and their other assumptions, CMC follows without assuming MOD*, so no argument is at hand that MOD* supports CMC. MOD* is supposed to replace CM1 and provide an independent basis for CMC. Even if the proof were valid, I do not think that this would be a very sensible basis since, as I have just been arguing

Conclusion 5: MOD* is highly implausible unless dependencies between causally unconnected quantities are already ruled out in the first place.

9.6   Two alternative claims and their defects

Let us try some other formulations of Hausman and Woodward's claim to see if they fare better. For their proof they need MOD* to hold in the specific population in which the parents of X_i take a fixed value. Perhaps then they intend that MOD* should hold in every population and hence in the requisite one:

H-W-claim$_2$: if MOD* holds for every Φ, $\mathcal{V}$, $\mathcal{U}$, G, P such that $\mathcal{V}$ is causally sufficient relative to G and P, then CMC holds as well for every Φ, $\mathcal{V}$, $\mathcal{U}$, G, P such that $\mathcal{V}$ is causally sufficient.

Given the antecedent, it is true that for any population, in the subpopulation where the parents of X_i take fixed values, $U_i' \perp X_j$; i.e. for every population, $U_i' \perp X_j/pa_i$. The consequent then follows that CMC holds for every population. Figure 9.1 is no longer a counterexample, since by inspection we can see

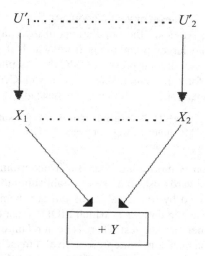

Figure 9.3 For subpopulation Φ'

that there is a population – the subpopulation of Φ picked out by fixing a value for Y – for which MOD* is violated; this is ruled out by the antecedent of the reformulated claim.

But the antecedent for this formulation is altogether too strong: it does not hold for a vast array of perfectly ordinary situations, including a host of ones in which CMC is satisfied. Consider, for example, a population Φ with probability measure P in which (where causes are on the right-hand side):

$$Y \leftrightarrow X_1 \text{ or } X_2$$
$$X_1 \leftrightarrow U'_1$$
$$X_2 \leftrightarrow U'_2$$
$$U'_1 \perp U'_2$$
$$\neg(X_1 c \rightarrow X_2)$$
$$P(U'_1) = P(U'_1/U'_2) = P(U'_1/U'_2 \& X_2 \& Y) = r \neq 1$$

where all the variables are dichotomous. For this population CMC holds.

Consider next a second population, Φ' – the subpopulation of Φ picked out by $+Y$. In this subpopulation U'_1 sets the value of X_1, but $P(U'_1/\neg X_2 \& + Y) = 1 \neq P(U'_1/X_2 \& + Y)$. So $\neg(U'_1 \perp X_2)$ in Φ', as illustrated in fig. 9.3.

Or look again at fig. 9.2 and consider the subpopulation in which $\neg X_1 \text{ v } X_2$. In this population U'_1 is still an intervention on X_1 and X_1 still does not cause X_2, yet $P(X_2/U'_1 = \text{on}) = 1 \neq P(X_2/U'_1 = \text{off})$.

It is, however, almost certain that Hausman and Woodward do not wish to formulate their claim in this way. After all, the populations in my examples are unrepresentative relative to the larger populations from which they are drawn, and we see by condition A that in their proof of CMC they assume that 'selection bias and other sources of unrepresentativeness' are absent. Certainly my subpopulations suffer from 'selection bias'. So let us try instead

H-W-claim$_3$: *if* MOD* holds for every Φ, $\mathcal{V}$, $\mathcal{U}$, G, P such that $\mathcal{V}$ is causally sufficient relative and Φ is representative, then CMC holds as well for every Φ, $\mathcal{V}$, $\mathcal{U}$, G, P such that $\mathcal{V}$ is causally sufficient and Φ is representative.

The antecedent in this formulation is more plausible. But it undermines the argument that Hausman and Woodward wish to make in establishing the consequent. The subpopulations selected by fixing values of pa_i are themselves unrepresentative, and it is just these populations in which MOD* must hold if CMC is to be deducible in the manner they suggest. There is a central unresolved issue about how to define 'selection bias' and 'unrepresentative'. I myself think that it is very difficult to do for purposes of defending CMC in general. In this case in particular I see no promise for defining it in a way that is not *ad hoc* and yet counts all unrepresentative subpopulations as biased except those selected by pa_i for each X_i in any variable set we may wish to consider.

I am thus led to

Conclusion 6: of the two alternative plausible readings, the first claim has a blatantly false premise and the second has no valid argument to support it.

9.7   A true claim and a valid argument

A more direct approach would be to formulate the thesis to say explicitly what is required for Hausman and Woodward's proof:

H-W-claim$_4$: for every Φ, $\mathcal{V}$, $\mathcal{U}$, G, P, if
  (i) for all X_i and all assignments of values, pa_{jk}, to the parents of X_i in G (Φ_{ik}, $\mathcal{V}$, $\mathcal{U}$, $G(pa_{ik})$, $P(pa_{ik})$ satisfies MOD*);
  (ii) $P(pa_{ik})(-) = P(-/pa_{ik})$ and $G(pa_{ik}) = G$;
  (iii) $\mathcal{V}$ is causally sufficient
then Φ, $\mathcal{V}$, $\mathcal{U}$, G, P satisfies CMC,

where $G(pa_{ik})$ is a graph of the causal relations over $\mathcal{V}+\mathcal{U}$ in the subpopulation of Φ in which the parents of X_i take the values pa_{ik}, and $P(pa_{ik})(-)$ is the probability distribution over $\mathcal{V}+\mathcal{U}$ in that same subpopulation.

H-W-claim$_4$ is true and the argument that Hausman and Woodward give in M&CMC shows that it is valid. But it does not gain Hausman and Woodward what they want – a route from manipulability/testability to CMC, for three reasons.

1 Claim$_1$ – the claim Hausman and Woodward seem to make, that any representative causally sufficient system that satisfies MOD* also satisfies CMC – is an interesting and surprising claim. Claim$_4$ is not. It tells us that if a very special set of unrepresentative populations, all subpopulations of Φ, satisfy MOD* then Φ will satisfy CMC. Now this may seem to be some gain, but I am afraid that it is very little, and it is certainly not the gain I had hoped for in connecting metaphysics and methods. For we have no more reason for accepting that the premise should be true of any given population than we would for expecting CMC in the first place. We could think that these very special populations satisfy MOD* because all populations do, and they do because manipulability is essential to causality. We have seen that that does not work because MOD* rests on testability not manipulability.

So what if we were to suppose that testability in the form MOD* itself is essential to causality? This still gets us nowhere in the proof for it takes us back to claim$_3$, and we have seen the problems with the premise in claim$_3$. It is false that MOD* holds for all populations: the unrepresentative one picked out by $+Y$ in fig. 9.3 shows that. But if the premise is restricted to all representative populations, the argument does not go through. For a valid proof we need to suppose that MOD* applies for just the right special set of unrepresentative populations. I do not find any independent reason for that in Hausman and Woodward's discussions.

2 The problem pointed out in section 9.6 still arises. $\mathcal{V}$ is causally sufficient but we do not presume from this that the U'''s are independent. Nor do we suppose CM1 to ensure they are independent. That is, they are not independent because they are causally unconnected – that it seems is not enough. But when we add that they set the values for quantities represented in $\mathcal{V}$, that is enough. But why?

3 The claim does not after all connect testability with CMC. Rather, it lays down very strong constraints on the populations, variable sets and graphs for which CMC is derived, and these constraints are strong enough to ensure both testability and CMC. This is exactly the same kind of problem that beset their earlier proof. We have a set of constraints C; C implies testability and C implies CMC. Of course by logic then, in C, testability implies CMC. But that is because in C, anything implies CMC. It is the constraints that imply CMC, not testability. In this case the constraints are conditions (i)–(iii) in the antecedent of H-W-claim$_4$.

But isn't constraint (i) itself an assertion of testability, and the inclusion of constraint (i) is essential to the truth of claim$_4$, as we all admit? No, constraint (i) is not a reasonable assertion of testability: it guarantees testability, but is itself stronger and stronger in just the way necessary to guarantee CMC.

Here is what I would take instead to be a reasonable statement of testability:

$\mathcal{V}$ is c-testable in Φ relative to $\mathcal{U}, G, P$ iff for all X_i in $\mathcal{V}$, there is an intervention Z_i in $\mathcal{V} + \mathcal{U}$ such that for all $X_j[\neg(X_ic \rightarrow X_j) \rightarrow P(X_j/Zi = \text{on}) = P(X_j/Zi = \text{off})]$.

I call this c-testability to stress that it is only one kind of testability – the kind we identify with a controlled test. As discussed in section 9.4, I myself would want to make the condition on the probabilities both necessary and sufficient for testability; but I do not do so here in order to stay as close as possible to Hausman and Woodward's formulations.

Notice how c-testability for $\Phi, \mathcal{V}, \mathcal{U}, G, P$ differs from MOD*. In the first place, c-testability requires that there be an intervention for every variable. On the other hand, it does not require that everything that counts as an intervention on Hausman and Woodward's definition should satisfy the independence assumption, merely that each variable has an intervention that does so. Hence nothing about c-testability automatically forces the U''s to satisfy the requisite independence assumption. This is for the reasons I have rehearsed. In Hausman and Woodward's scheme, we do not assume that a factor's being causally unconnected with others in the right ways is sufficient for guaranteeing the independence assumption; adding that that factor sets the value for a variable in $\mathcal{V}$ does not seem to add any reason for it to do so. On the other hand, if there is such a factor for each variable, then any hypothesis about one variable in $\mathcal{V}$ causing another can be tested.[23]

The point now is that MOD* (i.e. (i)), (ii) and (iii) guarantee c-testability as well as CMC. But c-testability in combination with (ii) and (iii) does not guarantee MOD* or CMC. That is because testability does not require that intervention be via a U' – it just requires there to be some intervention for each variable, and that is compatible with the U''s not being mutually independent. Figure 9.4 shows a particularly simple case:

The equations for the population Φ in fig. 9.4 are

$$X_1 c = U_{11} \vee U_{12}$$
$$X_2 c = U_{21} \vee U_{22}$$

with $P(i_{11}U_{11}, i_{12}U_{12}, i_{21}U_{21}, i_{22}U_{22}) = P(i_{11}U_{11})P(i_{21}U_{21})P(i_{22}U_{22})$ if $i_{12} = i_{21}$, and otherwise, where $i_{jk} = +, \neg$ and $P(i_{12}U_{12}) = P(i_{21}U_{21})$.

In this case U_{11} is an intervention on X_1: conditions (i)–(iv) in the definition of intervention are met by inspection and if U_1 occurs – call that 'on' – X_1 occurs no matter what values other variables take, so (v) is met as well. Similarly U_{22} is an intervention on X_2. Also, $U_{11} \vee U_{12}$ is an intervention on X_1 and $U_{21} \vee U_{22}$ is an intervention on X_2. $P(X_2/U_{11}$ is on$) = P(X_2/U_{11}$ is off$)$

---

[23] We should make special note of this last as well, for it is a very strong notion of testability – we want to be able to test every single causal hypothesis about the variables in $\mathcal{V}$.

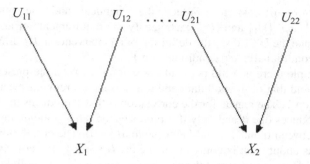

Figure 9.4

and $P(X_1/U_{22}$ is on$) = P(X_1/U_{22}$ is off$)$. But $P(X_2/U_{11} \vee U_{12} = $ on$) \neq$ $P(X_2/U_{11} \vee U_{12} = $ off$)$ and $P(X_1/U_{21} \vee U_{22} = $ on$) \neq P(X_1/U_{21} \vee U_{22} = $ off$)$. So $\mathcal{V} = \{X_1, X_2\}$ is c-testable in Φ relative to $\mathcal{U} = \{U_{ij}\}$, G, P. Conditions (ii) and (iii) of the antecedent of H-W-claim$_4$ are met as well. But condition (i) of that claim is not met and correlatively, CMC fails. C-testability obtains without the strong assumption needed for the true H-W-claim and without CMC.

The claim I have formulated as H-W-claim$_4$ is the only one I have been able to construct that makes their basic argument valid. If I am right that that is the only claim supported by their argument, then . . .

Conclusion 7: Hausman and Woodward can, using their basic ideas, produce a true claim and a valid argument. But their argument does not show that testability implies CMC; rather the constraints they need imply both testability and CMC; without these constraints, c-testability does not imply CMC.

9.8   Indeterminism

So far I have discussed only the deterministic case. For indeterminism we need more because in the probabilistic case a cause may produce a product and a by-product – i.e. two effects in correlation – and in this case the causal Markov condition will be violated. I have suggested for instance that a factory might produce an unwanted pollutant as a side effect during a purely probabilistic process that produces a desired chemical. In my comments on Hausman and Woodward's proof I represented this example thus:

$$X_1 \ c^= \ \alpha_1 Y + U_1$$
$$X_2 \ c^= \ \alpha_2 Y + U_2$$
$$P(+\alpha_1) = .8 = P(+\alpha_2)$$

Here Y is the presence of the chemical factory process; X_1, the presence of the chemical; X_2 the presence of the pollutant; α_1 and α_2 the operation of the chemical factory process to produce the chemical and the pollutant respectively;[24] and '$[U_1]$ and $[U_2]$ each satisfy the requirements of an intervention'.[25] Since the U's satisfy the definition of an intervention, $U_1 \perp X_2$ and $U_2 \perp X_1$, unconditionally and conditional on Y.

In the example there is a 100 per cent correlation between the presence of the chemical and that of the pollutant and this correlation remains even when we condition on Y. The reason for the correlation is that Y produces the two in tandem; it produces one if and only if it produces the other (though any other correlation between 0 and 1 could be possible as well). The correlation need not confuse us about what is going on. Since the U's satisfy the criteria of an intervention, it is easy to test that the chemical is not causing the pollutant, nor the reverse; and supposing that Y can be intervened on as well, it is easy to test that the chemical process is causing both.

Hausman and Woodward maintain that this kind of case is impossible, at least at the macro level. The issue is about $P(\alpha_1 \alpha_2)$. Can it, for instance, equal $P(\alpha_1)$, so that the pollutant is a byproduct of the chemical – it is produced iff the chemical is produced? If causation must be deterministic, this can easily happen but then CMC will not be violated because all the relevant probabilities will equal one. But we had best not assume that causality must be deterministic or we will not be able to say that what causes us to see the stars is the emission of photons that occurred on them long ago. So what happens when causation is probabilistic?

Hausman and Woodward maintain that it is impossible in this case for a cause to produce its effects together – it must produce one effect independently of the other. They argue that this is assumed on all standard accounts of causation. I do not agree. What kinds of thing do we expect of causation in our various standard accounts? Here are a few: (a) causes should make their effects happen. Y does that for both X_1 and X_2. (b) In the nice cases where all probabilistic dependencies can be derived from the causal laws operating, MOD* should be

[24] We need not be distracted about the issue of whether or not when an effect follows the occurrence of a purely probabilistic cause we should think that there is an additional event of the cause's 'firing' or 'producing' the effect. If we do not want to admit these kinds of events, we can take the α's to be mere notational devices that allow us to represent causal claims-cum-probability distributions as equations.

[25] Cartwright (2002), p. 436. The requirements for an intervention are slightly different in the new paper from any versions in the old. For the definition in the new paper I am not quite sure how they envisage writing equations where some of the U's are interventions. Perhaps $Y \, c= \delta_Z(\Sigma a_i X_i + U_y) + y^*Z$ for some chosen value y^* of Y, where $\delta_Z = 1$ when $Z \varepsilon$ $\mathcal{U} = 0$ (i.e. Z = off) and $\delta_Z = 0$ when $Z = 1$. The exact formulation does not matter though, since I began my formulation with a perfectly standard deterministic case where the U's satisfied the requirements for an intervention, whatever Hausman and Woodward wanted these requirements to be, then simply changed the operation of the factory from one that produced the chemical and the pollutant deterministically to one that produced them probabilistically, leaving intact from the previous deterministic case any alternative factors that can intervene and create the chemical or the pollutant independently of the action of any other causes.

satisfied. And it is. (c) In many situations if we put a mark on the cause we expect to find a trace of the mark on the effect. There is no reason to think that we cannot mark Y and find a trace later on both X_1 and X_2. So causation in this case has a great many of the features we expect of it.

If causes can produce their effects in tandem, CMC is violated. To prove CMC, Hausman and Woodward rule this possibility out directly with a premise they call 'no spontaneous correlation':

for every Φ, $\mathcal{V}$, $\mathcal{U}$, G, P and for every X_j ε $\mathcal{V}$ distinct from X_i, if $X_j \perp U_i'$, then $X_j \perp X_i/pa_i$.

I of course reject this premise. I also think the name may be misleading. The correlations that remain between X_1 and X_2 given Y's occurrence do not arise 'spontaneously' in the same sense in which time trends do or Sober's correlations between Venetian sea levels and British bread prices. They arise from the occurrence of a cause and the way it operates.

This brings us to one of the nice features of Hausman and Woodward's proof. They make very clear that even for causally sufficient variable sets, CMC could be violated for two different reasons: 'brute' dependencies not following from the causal principles governing the system as with time trends and bread prices and those due to causes producing their effects in tandem. They then offer separate cures for each: MOD* for the first, no spontaneous correlations for the second. This is a strong point about their proof – this distinction is clearly drawn and the separate problems are ruled out by separate premises. As they intended, it makes it easy to see where disagreements lie. I clearly reject the second of these premises.

What about the first? Here I take issue with Hausman and Woodward's discussion of my view. They spend a great deal of effort in reconstructing the factory example exactly as I presented it in my comments on their first proof. They then say, 'to the extent to which Cartwright is unwilling to commit herself to specific claims about what would happen under various interventions . . . it seems to us she has not clearly specified the causal structure of the example'.[26] But it is clear from the formulation what happens: intervene by manipulating U_1 and X_1 changes because U_1 causes X_1; X_2 and $P(X_2)$ do not change because U_2 and Y cause X_2 and since U_1 is an intervention, changes in it are supposed

---

[26] Sometimes I think Hausman and Woodward conflate the issue of whether there are interventions (as defined in any of the ways they propose) that can set the values (or probabilities) of the chemical and pollutant independently of what other causes for them are doing with the question of whether it is possible to stop Y itself from causing X_1 without stopping it from causing X_2. The formulation I gave is explicit about the first – which is what matters for MOD* and for tests of whether, for example, the chemical causes the pollutant or not (i.e. in their language, for 'disentangling' the common cause explanation of the correlation between chemical and pollutant from a direct cause account), but my formulation is silent about the second. The answer would presumably differ from one case to another, depending on the facts of the situation.

not to change U_2 and Y since they are not effects of X_1; $P(X_2/Y)$ does not change because $U_1 \perp X_2$, unconditionally and conditional on Y; and of course $P(X_2/X_1)$ does change.

Hausman and Woodward also say 'Cartwright's case that the chemical factory example is a genuine counterexample to [CMC] seems most plausible if one accepts MOD*',[27] suggesting by this and other remarks that I do not. On the contrary, I accept MOD* for a vast array of cases[28] and I built the chemical factory formulation to satisfy it. As they say, we must be assuming MOD* or something like it every time we draw a causal conclusion from a controlled experiment.

They also take issue with me for accepting in the case of the chemical factory that 'It should make no difference to the value of $[X_1]$ whether we set $[X_2]$ [by intervention] or observe $[X_2]$ once we set the parents of $[X_1]$ [i.e. once we set Y by intervention]'[29] while rejecting their claim called PM2 as it applies in the chemical factory case; PM2: $P(X_1/\text{set-}Y \,\&\, X_2) = P(X_1/\text{set-}Y \,\&\, \text{set-}X_2)$. But it is right to accept the first for the chemical factory example and reject the second.

Imagine an occasion on which we set Y so that Y must occur. Y occurs. On this occasion Y produces X_1 and thus, since Y produces X_1 iff Y produces X_2, X_2 occurs. If we also on this occasion intervene on X_2 to make X_2 occur, X_2 will still occur – it will be overdetermined – and so will X_1 occur. So whether we intervene on X_2 will make no difference to the value of X_1. Imagine on the other hand that Y does not produce X_2, so X_1 does not occur on this occasion. If we were to produce X_2 by intervening, that will not make Y suddenly produce X_1 so X_1 will still not occur. Again, whether we intervene on X_2 will make no difference to the value of X_1.

But the claim about probabilities does not follow from the claim about values and is indeed false. The conditional probabilities of X_1 change although the values never do for the usual reason. Imagine Y is set. Then when the intervention is off, all X_2 occasions will be X_1 occasions. But among the set-X_2 occasions, only 80 per cent will be X_1 occasions; that is true just because no $\neg X_1$ occasion ever turns into an X_1 occasion just by turning the occasion from a $\neg X_2$ one into an X_2 one.

They also say that I cannot endorse the first claim and accept the arrow-breaking interpretation of intervention that they offer in their new proof and that I suppose in my chemical factory case. But that is a mistake too. Perhaps Hausman and Woodward think that intervening on X_2 will interfere with Y's operations, but obviously that should not be the case for an intervention. Setting $U_2 = 1$

[27] Hausman and Woodward (2004), p. 159.
[28] Though not all cases. I think brute correlations may well occur in many situations; we want to be sure they aren't happening whenever we draw causal conclusions from correlations.
[29] Hausman and Woodward (2003) ms p. 19, fn. 11. This is an earlier version of Hausman and Woodward (2004).

should leave Y's operations unaffected. (Here we see some of the complications in defining 'intervention' – obviously in cases of probabilistic causality we want to ensure that an intervention on one variable does not interfere with whether another would or would not produce its result on any occasion.)

In their discussion of product/by-product cases, Hausman and Woodward argue that 'the explication of causal claims in terms of what would happen under various hypothetical interventions does provide . . . an independent purchase [on the content of causal claims]'.[30] I agree that it does – so too do all the other theories of causation on offer and all the other methods (like the mark method) that we use to test for causality. But even if we took theirs as the central purchase, it does not help the case for CMC nor provide support for the no-spontaneous-correlation premise since MOD* can be readily satisfied in cases where causes produce their effects in tandem.[31]

So I draw

Conclusion 8: product/by-product cases that violate CMC can be ruled out by a specially designed premise but that does not show much. And it is no help in establishing a route from testability to CMC.

9.9   Overall conclusion

The route from manipulability/testability to CMC is not there. CMC is not a reflection of any important metaphysical facts about causation. And anyway, those putative facts about causation are not facts!

---

[30] Hausman and Woodward (2004), p. 159.

[31] Hausman and Woodward also, in passing, try to defend the view that it should be possible to manipulate each factor separately – that is, that intervention is always possible. They do so by attacking my claim that equations that provide information about a full set of causes need not also provide information about what can and cannot be manipulated separately. Their argument is just their argument in favour of MOD* – 'in the absence of modularity there will be changes in the values of variables under interventions on other variables that are not reflected in the causal claims expressed in the system of equations' (Hausman and Woodward 2004, p. 158). This argument is invalid since the premise supports MOD*, which states what happens if intervention occurs, but the conclusion is that intervention is always possible.

10 Two theorems on invariance and causality

10.1 Introduction

10.1.1 The project

Much recent work on causal inference takes invariance under intervention as a mark of correctness in a causal law-claim.[1] Often this thesis is simply assumed; when it is argued for, generally the arguments are of a broad philosophical nature with heavy reliance on examples. Also, the notions involved are often characterized only loosely, or very specific formulations are assumed for the purposes of a particular investigation without attention to a more general definition, or different senses are mixed together as if it did not matter. But it does matter because a number of different senses appear in the literature for each of the concepts involved, and the thesis is false if the concepts are lined up in the wrong way.

To get clear about whether invariance under intervention is or is not necessary or sufficient for a causal-law claim to be correct, and under what conditions, we need to know what counts as an intervention, what invariance is, and what it is for a causal-law claim to be correct. Next we should like some arguments that establish clear results one way or the other. In this chapter I offer explicit definitions for two different versions of each of the three central notions: intervention, invariance and causal claim. All of these different senses are common in the literature. Then, given some natural and relatively uncontroversial assumptions, I prove two distinct sets of theorems showing that invariance is a mark of causality when the concepts are appropriately interpreted. These, though, are just a sample of results that should be considered.

Thanks to Daniel Hausman and James Woodward for setting me off on this project and two referees for helpful suggestions. This research was funded by a grant from the Latsis Foundation, for which I am grateful, and it was conducted in conjunction with the Measurement in Physics and Economics Project at LSE. I wish to thank the members of that group for their help, especially Sang Wook Yi and Roman Frigg.
[1] Glymour, Scheines, Spirtes and Kelly (1987); Hausman and Woodward (1999); Hoover (2001); Redhead (1987); Woodward (2003). See also the discussion of David Hendry in ch. 16 of this book.

The two different sets of theorems use different senses of each of the three concepts involved and hence make different claims. Both might loosely be rendered as the thesis that a certain kind of true relation will be invariant when interventions occur. In the second, however, what counts as 'invariance' becomes so stretched that the term no longer seems a natural one, despite the fact that this is how it is sometimes discussed in the literature – especially by James Woodward, whose extensive study of invariance is chiefly responsible for isolating this particular characteristic and focusing our attention on it.

Nor is 'intervention' a particularly good label either. The literature on causation and invariance is often connected with the move to place manipulation at the heart of our concept of causation:[2] roughly, part of what it means to be a cause is that manipulating a cause is a good way to produce changes in its effects. 'Manipulation' here I take it suggests setting the target feature where we wish it to be, or at will or arbitrarily. Often when authors talk about intervention, it sounds as if they assume just this aspect of manipulation.

Neither set of theorems requires a notion so strong. All that is required is that nature allow specific kinds of variation in the features under study.[3] We might argue that manipulability of the right sort will go a good way towards ensuring the requisite kind of variability. But mere variation of the right kind will be sufficient as well, so we need to take care that formulations employing the terms 'manipulation' and 'intervention' do not mislead us into demanding stronger tests for causality than are needed.

In this chapter I am concerned only with claims about deterministic systems where the underlying causal laws are given by linear equations linking the size of the effect with the sizes of the causes. Although this is extremely restrictive, it is not an unusual restriction in the literature, and it will be good to have some clean results for this well-known case. The next step is to do the same with different invariance and intervention concepts geared to more general kinds of causal systems and less restrictive kinds of causal-law claims.

This project is important to practising science. When we know necessary or sufficient conditions for a causal-law claim to be correct, we can put them to use to devise real tests for scientific hypotheses. And here we cannot afford to be sloppy. Different kinds of intervention and invariance lead to different kinds of tests, and different kinds of causal claims license different things we can do. So getting the definitions and the results straight matters to what we can do in the world and how reliable our efforts will be.

---

[2] Price (1991); Hausman (1998); Woodward (1997); Hausman and Woodward (1999); Woodward (2003).

[3] Or, if the right kind of variation does not actually occur, there must be a fact of the matter about what would happen were it to do so.

10.1.2   The nature of deterministic causal systems

I need in what follows to distinguish between causal laws and our representations of them; I shall use the term 'causal system' for the former, 'causal structure' for the latter. I take it that the notion of a 'causal law' cannot be reduced to any non-modal notions. So I start from the assumption that there is a difference between functional relations that are just true and ones that are true in a special way; the latter are nature's causal laws. I will also assume transitivity of causal laws. This implies that the causal systems under study include not only facts about what causal laws are true – e.g. 'Q causes P' – but also about the possible ways by which one factor can cause another – e.g. 'Q causes P via R and S but not via T'.

I discuss only linear systems, and I shall represent nature's causal equations like this: q_e c$= \Sigma a_{ej} q_j$, with the effect on the left and causes on the right. As will be clear from axiom A_1, this law implies that $q_e = \Sigma a_{ej} q_j$ but not the reverse. Following the distinction between systems and structures, I shall throughout use q_i to stand for quantities in nature and x_i for the variables used to represent them. Also with respect to notation, I shall use lower case letters for variables and quantities and upper case letters for their values. I assume the following about nature's causal systems:

A_1 Functional dependence: any causal equation presents a true functional relation.

A_2 Anti-symmetry and irreflexivity: if q causes r, r does not cause q.

A_3 Uniqueness of coefficients: no effect has more than one expansion in the same set of causes.

A_4 Numerical transitivity: causally correct equations remain causally correct if we substitute for any right-hand-side factor any function in its causes that is among nature's causal laws. (For a discussion of this axiom, see footnote 3 in ch. 5 in this book.)

A_5 Consistency: any two causally correct equations for the same effect can be brought into the same form by substituting for right-hand-side factors in them functions of the causes of those factors given in nature's causal laws.

A_6 The priority of causal relations: no quantities are functionally related unless the relation follows from nature's causal laws.

More formally: a linear deterministic system (LDS) is an ordered pair $\langle Q, CL \rangle$, where the first member of the pair is an ordered set of quantities $\langle q_1, \ldots, q_m \rangle$ and the second is a set of causal laws of the form q_k c$= \sum_{j<k} a_{kj} q_j$ (a_{kj} a real number) that satisfies A_1 through A_6.[4]

---

[4] More precisely a causal law for an effect x_j, $L(x_j)$, is a set of ordered pairs giving causes of x_j and their weights $L(x_j) = \{\langle a(1)_{j1}, x_1 \rangle \langle a(2)_{j1}, x_1 \rangle \ldots \langle a(k_1)_{j1}, x_1 \rangle \langle a(1)_{j2}, x_2 \rangle \ldots \langle a(k_n)_{jn}, x_n \rangle\}$, $a(k)_{jm} \varepsilon R$. We can then define x_i causes x_j with weight a just in case $\exists L(x_j)(\langle a, x_i \rangle \varepsilon L(x_j))$. (Notice that my formulations allows – as I have argued we should – for a cause to have multiple

10.2   Causal law variation, invariance and one kind of causal claim

10.2.1   The first definitions

The kind of intervention we shall be concerned with in this section is the same as that employed by Pearl in his work on causal counterfactuals and by Glymour et al. in their manipulation theorem (once we transform their notion from graph representations to linear deterministic systems). It is also one of the kinds that Daniel Hausman and James Woodward discuss in their joint work on the Markov condition.[5]

As I indicated in section 10.1, the results I aim to establish are not really results about intervention in the natural sense of that term, but rather results about variation. The first kind of intervention, which will be under discussion here in section 10.2, is one in which causal laws vary; in the second kind, which I discuss in section 10.3, it is the values of the causes picked out in a fixed causal system that vary. We may perhaps be more used to thinking of quantities as taking on different values than of laws as varying.[6] But all we need here is that there are different causal systems that relate to each other in the specific way I shall describe. The point I am trying to make is that it is the occurrence of these systems that matters[7] for testing the correctness of causal claims; it is not necessary that they come to occur through anything naturally labelled an intervention or a manipulation.[8] I shall, therefore, talk not of intervention but rather of variation.

---

capacities with respect to the same effect. Once we have admitted this piece of information we can of course go on to define some concept of 'the overall influence' of a given cause on a given effect.)

　　Clearly the assumptions too need a more precise formulation. Transitivity, for example, becomes

A$'$4 For any laws $L(x_j)$ and $L(x_i)$, and for any $\langle b, x_i \rangle \varepsilon L(x_j)$, $L'(x_j)$ is also a law, where

$$L'(x_j) = L(x_j) - \{\langle b, x_i \rangle\}\{\langle (ba'(1)_{i1}, x_1\rangle, \ldots, \langle (ba'(k_n)_{in}, x_n\rangle\} \text{ for all } \langle a'(k_m)_{im}\rangle \varepsilon L(x_i)$$

The other assumptions are formulated similarly.

　　We need some kind of complicated formulation like this to make clear, e.g. that arbitrary regroupings on the right-hand side of a causal-law equation will not result in a causal law. For example, assume that x_2 c$^=$ ax_1 and x_3 c$^=$ $bx_1 + cx_2$. It follows that $x_3 = bx_1 + (c - d)x_2 + dx_2 = bx_1 + (c - d)ax_1 + dx_2 = (b + ca - da)x_1 + dx_2$, but we do not wish to allow that x_3 c$^=$ $(b + ca - da)x_1 + dx_2$. For our purpose here, I think we can proceed with the more intuitive formulations in the text.

[5] Pearl (2000b); Glymour et al. (1987); Hausman and Woodward (1999).

[6] In my own work (Cartwright (1999)) on laws it is natural that they should vary since laws are epiphenomena, depending upon stable arrangements of capacities. I take the prevalence of 'intervention' tests for causal correctness of the kind described here, based on the possibility of variations in causal laws, to indicate that a surprising number of other philosophers are committed to something like my view.

[7] Or, the possibility of the occurrence of these systems (see n. 3).

[8] There are of course other kinds of arguments for linking manipulation and causation (e.g. Hausman (1998); Price (1991)). My point here is that it is mistaken to argue that manipulation is central to causation on the grounds that one important kind of test for causal correctness – the 'invariance' test – cannot do without it.

In the first kind of 'variation'/'intervention', which I call causal-law variation, a new causal system is considered, similar in many ways to the first. Let us call the new system a test system for results of quantity q relative to the original system. The test system differs from the original that we wish to test by exactly one addition and two kinds of deletions. For a target quantity q, add the law $q = Q$ for some specific value, Q, of q within its allowed range. Drop (1) all laws with q as effect and (2) all laws linking causes of q with effects, e, of q where the causal influence passes through q – that is, any equation for e that can be obtained by transitivity from an equation giving q's effects on e. The first is easy to say formally: drop all laws of the form $q \text{ c}^= f(\ldots)$. The second is more cumbersome: drop any equation A: $e \text{ c}^= f(\ldots, g(\ldots), \ldots)$ for which there are equations of the form B: $e \text{ c}^= f(\ldots, q, \ldots)$ and C: $q \text{ c}^= g(\ldots)$.

As with 'intervention', there are a number of different kinds of invariance suggested in the literature. The one relevant here seems genuinely a notion of invariance, so that is what I shall call it. An equation in a (linear deterministic) causal system $\langle Q, CL \rangle$ giving a true functional relation (but not necessarily one that replicates one of nature's causal laws) is invariant in q iff it continues to give a true functional relation for any value that q takes in any test situation for q relative to $\langle Q, CL \rangle$.

We also need to be explicit about what an equation of the form $x_e \text{ c}^= \Sigma a_i x_i$ in a causal representation is supposed to be claiming. I propose the obvious answer: an equation of this form records one of nature's causal laws. When it does so, I shall say that it is causally correct.

10.2.2   The first theorem

A functionally true equation is causally correct iff it is invariant in all its independent variables, either singly or in any combination.

Correctness → invariance: the result in this direction is trivial now that the background is in place. Consider an equation that is causally correct:

$$E : x_e \text{ c}^= f(x_1, \ldots, x_n)$$

Consider a test system for the effects of q_i for any q_i represented by an x_i in the right-hand side of E. The intervention replaces the causal system of which this equation is a part by a new one. This equation would be dropped from the new system if it had q_i as an effect – which it hasn't. Otherwise it would be dropped only if it had as effect an effect of q_i – which it has – and it results from substituting $g(\ldots)$ for q_i into some equation for q_e, where $q_i \text{ c}^= g(\ldots)$. But in this case q_i would no longer appear in the equation to be dropped. $x_e \text{ c}^= f(x_1, \ldots, x_i, \ldots, x_n)$ will still obtain in the new system. Hence E is invariant under interventions on q_i.

Clearly the trick in establishing the necessity of invariance for correctness is in the characterization of interventions. So we shall need to be wary when we introduce a different concept of intervention, as in section 10.3.

Invariance → correctness: consider an equation

$$F : x_e = \sum_{i=1}^{N} a_i x_i$$

where either some x_i appears that is not the cause of x_e, or, if all are genuine causes, some x_i appears with a causally incorrect coefficient. In order to be invariant, F must also be derivable in all test systems for all q_i and it must be derivable from the same equations as in the original. That is because the move to a test system adds only one kind of new law to use in a derivation – '$q_i = Q_i$' where Q_i may be any value in the appropriate range. This clearly will not help since Q_i will vary from test system to test system, and F must be derivable in all of them. But if F is derivable from the same set of laws in the test situation as in the original, then not only will F be invariant in all x_i, so too must each member of this set be. So we wish to establish:

No matter what the causal system, no linear combination of nature's causal equations will yield an equation of form F that is invariant in all the q_i represented on the right-hand side of F.

We should first notice that, trivially,

Claim 1: no matter what the causal system, no causal equations used in the linear combination can have an x_i on the left-hand side.

The result is then established by coupling claim 1 with:

Claim 2: no matter what the causal system, no linear combination of causal equations in which x_i's appear only on the right-hand side will yield F.

Proof of claim 2 The proof of claim 2 is by induction on the number of variables in addition to x_e and the x_i's that appear in the equations in the linear combination that yields F.

Inductive base: as a base for the induction, show that no linear combination of equations in any causal system that use no variables in addition to x_e and the x_i's and are invariant in all x_i will yield F. Here is how.

All equations used in such a linear combination will have x_e on the left-hand side and some combination of x_i's on the right-hand side. That is, they will look like this:

$$B : x_e \mathrel{c}^= \sum b_i x_i$$
$$C : x_e \mathrel{c}^= \sum c_i x_i$$
$$\vdots$$

where some of the b_i and some of the c_i will be zero. By consistency, some combination of factors from B cause factors in C or the reverse or both. But if factors in B cause a factor represented by[9] x_i in C, then B will not be invariant in x_i. Similarly, if factors in C cause a factor, x_i', in B, then C will not be invariant in x_i'. So no two such equations can be used and F cannot be so obtained.

Inductive argument: we aim to establish by *reductio* that if claim 2 is true for a set of equations using n variables in addition to x_e and the x_i's, it will be true for a set using $n + 1$ additional variables. So suppose F can be obtained using $n + 1$ additional variables; let $z_1, \ldots, z_k, k = N + n + 1$, denote the variables that appear in a linear combination that yields F.

Lemma. At least one of the 'extra variables' – one of the z_i that is neither x_e nor any of the x_i's – must appear as an effect in the equations used at least once. Call it z.

This is true because

(i) Among extra variables that appear as causes, at least one will not be a cause of any of the other extra variables involved. Otherwise we would have a causal loop, which violates anti-symmetry. Call it z'.

(ii) Since z' does not appear in F, it must appear in at least two equations (one to introduce it, one to eliminate it).

(iii) Both these equations must have x_e as effect since no x_i can appear as an effect in an invariant equation. z' could appear with the same coefficient in both equations:

$$x_e = az' + \sum a_i z_i$$
$$x_e = az' + \sum b_j z_j$$

By consistency, $\sum a_i z_i$ and $\sum b_j z_j$ can be brought into the same form by a set of laws, L, linking the z_i and the z_j. In this case these two equations containing z' can be replaced in F by the laws in L, which do not contain z', with no loss. Alternatively, z' can appear with different coefficients in the two equations:

$$x_e = az' + \sum a_i z_i$$
$$x_e = bz' + \sum b_j z_j$$

But this is possible only if z' is a cause of either one or more of the z_i or of the z_j.

Since these effects must be x_i's, the equation with the causes of these x_i's will not be invariant in all x_i.

---

[9] I shall henceforth drop the use of 'represented by' where it will not cause confusion and simply talk of variables causing other variables.

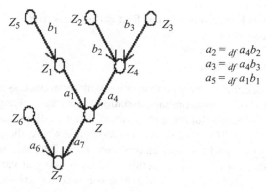

$$a_2 =_{df} a_4 b_2$$
$$a_3 =_{df} a_4 b_3$$
$$a_5 =_{df} a_1 b_1$$

Figure 10.1

We can now eliminate z in the following way. Consider nature's causal law for z as effect that cites as causes just those factors that are direct causes of z among the z_i. Designate it thus:

$$z \overset{=}{c} \sum a_i y_i \qquad y_i \in \{z_i, \ldots, z_k\}$$

Replace any equation in the original linear combination in which z appears as cause by the same equation with $\sum a_i y_i$ substituted for z. Eliminate all equations with z as effect.

Add nature's causal equations giving the relations between all the causes that appear in all the different equations that had z as effect, as well as those connecting z's parents with the effects of z among the z_i. For example, supposing the relations in Fig. 10.1, we replace

$$z \overset{=}{c} a_1 z_1 + a_2 z_2 + a_3 z_3$$
$$z \overset{=}{c} a_4 z_4 + a_5 z_5$$
$$z_7 \overset{=}{c} a_6 z_6 + a_7 z$$

with

$$z_1 \overset{=}{c} b_1 z_5$$
$$z_4 \overset{=}{c} b_2 z_2 + b_3 z_3$$
$$z_7 \overset{=}{c} a_6 z_6 + a_7 (a_1 z_1 + a_4 z_4)$$

Clearly the new set of equations will be invariant in all x_i if the original are, and any equation in x_e and the x_i that can be obtained using the original equations can be obtained using the new ones. QED.

10.3   Variation of values, prediction of first differences and parameter correctness

10.3.1   Systems that are nice for us

The basic idea in connecting intervention/variation with invariance as a test of causality is Mill's method of concomitant variation: as a cause changes, the effect should change 'in train'. But there are caveats. The variation must occur in the right circumstances. The easiest circumstances are where the putative cause varies all on its own and no other causes vary at all. That is essentially what we achieve in the test systems of section 10.2 by looking at variants of the original causal laws that make the putative cause take a particular value independent of what values other factors have.

But sometimes, if a causal system is sufficiently nice, we can achieve essentially the same results by looking within the system itself. The simplest case is where each of the putative causes for a given effect has a cause of its own that can vary without any cross-restraints on other possible causes of that effect. That will guarantee that all possible causes can take on any combination of values. I call such a system epistemically convenient.

More formally, an epistemically convenient linear deterministic system (ECLDS) is a linear deterministic system $\langle Q, CL \rangle$, such that

A_7 Epistemological convenience: for each q_j in $Q = \{q_1, \ldots, q_m\}$ there is some cause q_j^* such that:

(i) $q_j \overset{c}{=} \Sigma_{k<j} c_{jk} q_k + q_j^*$

(ii) there are no cross-restraints on the values of the q_j^*; that is, for all situations in which $\langle Q, CL \rangle$ obtains, it is possible ('allowed by nature') for each q_j^* to take any value in its allowed range consistent with all other q_k^* taking any values in their allowed ranges.[10]

In case the LDS we are studying is an epistemically convenient one, we can relabel the quantities so that the system takes the familiar form

$$q_1 \overset{c}{=} u_1$$
$$q_2 \overset{c}{=} a_{21} q_1 + u_2$$
$$\vdots$$
$$q_n \overset{c}{=} a_{n1} q_1 + \cdots + a_{nn-1} q_{n-1} + u_n,$$

---

[10] This is similar to a standard kind of condition on parameter values in econometrics (cf. Engle, Hendry and Richard (1983)) and as a condition on parameter values plays a central role in Kevin Hoover's (2001) theory of causal inference. Woodward (1997) asks for statistical independence of the exogenous quantities. The proof here requires the additional assumption that there are no cross restraints on their values.

where $n = m/2$. For the remainder of this part, I consider only epistemically convenient linear deterministic systems, and I assume that the notation has its natural interpretation for such systems.

Notice that (i) and (ii) imply

(iii) no q_k in Q causes q_j^*

but neither

(iv) for all j, k, q_j^* does not cause q_k^*

nor

(v) for all j, k, q_j^* and q_k^* have no common cause (i.e. they are not part of any other LDS in which they have a common cause).

Many authors restrict their attention to systems satisfying (iv) and (v) as well, usually with the intention of mounting an argument from (i), (iii), (iv) and (v) to (ii). I shall not do so because the argument is not straightforward and at any rate we need only the assumption (ii) to derive the results of interest here.

Following standard usage, let us call the 'special causes' represented by u's in an ECLDS, exogenous quantities, since they are not caused by any quantities in the system. Notice that, for an ECLDS, an assignment of values to each of the exogenous quantities will fix the value of all other quantities in the system. In what follows it will help to have an expression for a quantity in the system in terms of the exogenous quantities. Again following conventional usage, I call this the reduced form:

$$\text{RF}: \quad q_k \; \text{c}^= \; \sum_{i=1}^{k} u_i \sum_{l=i}^{k-1} a_{kl} \sum_{m=i}^{l-1} a_{lm} \ldots$$

where we adopt the convention

$$\sum_{j=\alpha}^{\beta} f_i(j, k, l, \ldots) = 1, \text{ if } \alpha > \beta$$

$$\therefore q_k \; \text{c}^= \; \sum_{i=1}^{k} \Gamma_i^k u_i$$

where $\Gamma_i^k = \sum_{l=i}^{k-1} a_{kl} \sum_{m=i}^{l-1} a_{lm} \ldots$

Call any set of values for each of the exogenous terms a situation. We shall be interested in differences, so let us define

$$\Delta_j^\alpha q_n =_{df} q_n(u_1 = U_1, \ldots, u_{j-1} = U_{j-1}, u_j = U_j + \alpha,$$
$$u_{j+1} = U_{j+1}, \ldots, u_{m/2} = U_{m/2}) - q_n(u_1 = U_1, \ldots,$$
$$u_{j-1} = U_{j-1}, u_j = U_j, u_{j+1} = U_{j+1}, \ldots, u_{m/2} = U_{m/2})$$

Statisticians like epistemologically convenient systems because they make estimation of probabilities from data easier. We, by contrast, are concerned with

how to infer causal claims given facts about association. For this project, these
kinds of systems have three advantages.

1 In section 10.2 we discussed methods for finding out about a causal system
of interest by looking at what happens in other related systems. But the
existence of the system of interest provides no guarantee that these other
systems exist for us to observe. In this part we shall be interested in situations
in which specified factors take arbitrary values relative to each other. In an
epistemologically convenient system this is guaranteed to happen 'naturally'
within the system itself – at least 'in the long run'.[11]

2 Consider a functionally correct hypothesis,

$$H : x_e \overset{c}{=} \Sigma a_{ej}x_j$$

where each q_j (represented by x_j) has an exogenous cause peculiar to it
satisfying (ii). In this case nature provides a basic arrangement that allows
the possibility for each q_j to have an open back path; whether indeed each
does have an open back path will depend entirely on our knowledge, but at
least the facts are right to allow us knowledge of the right kind. Relative to
q_e, q_j has an open back path just in case (a) every causal law with q_j as
effect has a u_j such that u_j cannot cause q_e except by causing q_j, and (b) we
know what these u's are and we know that (a) is true of them. (For further
discussion of open back paths see ch. 13 in this book.)

   The nice thing about hypotheses like H where every putative cause has an
open back path is that we can tell by inspection whether H is true or not. For
no x_j can appear in a functionally correct equation with a causally wrong
coefficient unless some factor appears on the right-hand side of that equation
along with a factor from its back path.[12] But according to (a), no factor from
the back path of q_j can appear as a cause of q_e in the same law as q_j. The
equation for x_e is thus a true causal law, so long as nothing appears on the
right-hand side that is from the back path of any other factor that appears
there. Given (b), we can tell this just by looking. According to Cartwright,[13]
J. L. Mackie's famous example of the London workers and the Manchester
hooters works in just this way.

3 Randomized treatment/control experiments are the gold standard for establishing causal laws in areas where we do not have sufficient knowledge to
control confounding factors directly. These experiments require that there be
some method for varying the causal factors under test without in any other
way producing variation in the effect in question. In an epistemologically

---

[11] Thanks to David Danks for highlighting this feature.
[12] The proof is similar to the proof of the theorems in section 10 above. See Cartwright (1989).
(Note that the argument in Spirtes, Glymour and Scheines (1993) against this result uses as a
putative counterexample one that does not meet the conditions set.)
[13] 1989.

convenient system, the exogenous quantities peculiar to each factor provide just such a method.

10.3.2   The second definitions

Now for 'intervening'. The idea is to 'vary' the value of the targeted quantity by adjusting its exogenous cause in just the right way keeping fixed the values of all the other exogenous causes. But as I indicated, neither the idea of our manipulating nor of our varying anything matters. All we need is to consider what would happen were two different values of the exogenous cause of the targeted quantity to occur in two otherwise identical situations. So I propose the following definition: a variation/intervention of values is a calculation of $\Delta_j^\alpha q_k$ for some j, k, α. Direct inspection of the reduced form for q_k shows the following to hold:

> Lemma (on reduced forms and causality): if q_j does not cause q_k then $\Delta_j^\alpha q_k = 0$.

Along with the notion of 'intervention', we have to introduce new notions of invariance and causal correctness as well, otherwise the kinds of theorem we are interested in will not follow. The result in one direction still follows: any causally correct equation will be invariant under variation/intervention. But that is because any true equation will be, including all those equations that suggest joint effects of a common cause as causes of each other. Hence the result we really want in order to test for causal correctness will not follow, i.e. it is not true that any equation that is invariant under value variation/intervention will be causally correct (even if we restrict attention, as below, to equations in which no right-hand-side quantity causes any other).

What notion shall we substitute for that of invariance? The answer must clearly be tied to what kind of causal claim is made since we are not, after all, interested in invariance itself but pursue it as a test for causality. So far the kind of causal claim we have considered is terrifically restricted, given our usual epistemic position. For we consider only hypotheses that claim to offer a complete (i.e. determining) set of causes and with exactly the weights nature assigns them. One way to be less demanding would be to ask for causes but not insist on weights, as in a Bayes-net graph.

Another alternative is to insist that the weights be correct, but not insist on a complete set of causes. This is the one I consider here. If we are offering claims with some causes omitted, what form should the hypotheses take? One standard answer is that they take the form of regression equations:

$$R: \quad x_k \, c^= \, \Sigma a_{kj} x_j + \Psi_k, \text{ for } \Psi_k \perp x_j \text{ for all } j$$

where $x \perp y$ means that $\langle xy \rangle = \langle x \rangle \langle y \rangle$. This of course only makes sense if there is a probability measure from which the expectations are derived. So the use of hypotheses of this form involves an additional restriction on the kinds of systems under study, as follows. An epistemically convenient linear deterministic system with probability measure (ECLDSwPM) is an epistemically convenient linear deterministic system that satisfies A_8.

A_8 Existence of a probability measure: the quantities in Q can be represented by random variables $x_1, \ldots, x_m$ which have a probability measure defined over them. (Following conventional notation, we can relabel the x's just as we have the q's so that $\{x_1, \ldots, x_m\} = \{x_1, \ldots, x_{m/2}, u_1, \ldots, u_{m/2}\}$).

What does an equation of form R assert? This kind of equation is often on offer but generally without any explanation about what claims it is supposed to make. I take it that it is supposed to include only genuine causes of x_k and moreover to tell us the correct weights of these. I propose, therefore, to define correctness thus: an equation of the form

$$R: x_k \stackrel{c}{=} \sum a_{kj} x_j + \Psi_k (1 \leq j \leq m/2)$$

for $\Psi_k \perp x_j$ is correct iff there exist $\{b_j\}$ (possibly $b_j = 0$), $\{q_j'\}$ such that

$$q_k \stackrel{c}{=} \sum a_{kj} q_j + \sum b_j q_j' + u_k (1 \leq j \leq m/2)$$

where q_j does not cause q_j'. This last restriction ensures that all omitted factors are causally antecedent to or 'simultaneous' with those mentioned in the regression formula.

It may be useful to consider an example:

$$q_1 \stackrel{c}{=} u_1$$
$$q_2 \stackrel{c}{=} a_{21} q_1 + u_2$$
$$q_3 \stackrel{c}{=} u_3$$
$$q_4 \stackrel{c}{=} a_{41} q_1 + a_{42} q_2 + a_{43} q_3 + u_4$$

In this causal system the equation

$$x_4 \stackrel{c}{=} (a_{41} + a_{42} a_{21}) q_1 + R$$

is correct. It may seem worrying that q_2 is omitted from the right-hand side of the regression equation and that it is caused by q_1, which is included. But this is alright. The claims of the regression equation are correct under the proposed definition because there is a true causal law in which the coefficient of q_1 is that given in the regression equation and no factors in the true law that do not appear in the regression equation are caused by ones that are mentioned.

Now return to the unresolved issue of what can be introduced in place of invariance to dovetail with this characterization of correctness for regression equations. As I indicated in the introduction to this chapter, the notion that I use is not a notion of invariance at all. It is rather a notion of correct prediction:

correct prediction of variation in values as situations vary in specific ways. This is not in any way a new notion, but it is one that Woodward has recently directed our attention to and that he has developed at length. I believe that what I define here is the right way to characterize his ideas when applied to epistemically convenient linear deterministic systems and I take it that the theorem I prove is one precise formulation of what he argues for (once a number of caveats are added to his claims).

What do equations of form R predict about the difference in the size of effect between these two situations? If R's claims are correct, the difference in the effect given a variation of the special exogenous variable that causes one of the right-hand-side variables, say x_J, should be thus: $\Delta_J^\alpha q_k = \Sigma a_{kj} \Delta_J^\alpha q_j + \Sigma b_j \Delta_J^\alpha q_j'$ for some $\{b_j\}$ and $\{q_j'\}$, where no q_j causes any q_j'. By inspection of the reduced form equations in an ECLDSwPM, we see that the second term on the right-hand side is zero, since q_J does not cause any of the quantities that appear there. So R's predictions are correct just in case $\Delta_J^\alpha q_k = \Sigma a_{kj} \Delta_J^\alpha q_j$. So let us define: an equation of form R correctly predicts first differences for all right-hand-side variables if and only if $\Delta_J^\alpha q_k = \Sigma a_{kj} \Delta_J^\alpha q_j$ for all α and for all J, where J ranges over the right-hand-side variables.

10.3.3   The second theorems

Now I can state the relevant theorem.

Theorem 2a   A regression equation for q_k, x_k c$= \Sigma_{j=1}^{k-1} a_{kj} x_j + \Psi_k$ is causally correct if for all α and for all J, $1 \le J \le k - 1$, $\Delta_J^\alpha q_k = \Sigma a_{kj} \Delta_J^\alpha q_j$; i.e. iff the equation predicts rightly the first differences in q_k generated from any value variation/intervention in any right-hand-side variable.

First a note on notation. In general there will be more q's in the underlying causal system than are represented by x's from the causal structure. For convenience I suppose that the q's are ordered following the x's: i.e. q_j is the quantity represented by x_j. This means that we cannot presuppose that q_i is causally prior to q_{i+l}.

Proof of theorem 2a   The proof from correctness to the prediction of first differences in q_k under variations of right-hand-side variables is trivial. To go the other direction, first reorder the q's so that they are numbered in their true causal order (so, q_j can only cause q_{j+l} for $l \ge 1$), which we can do without commitment since the ordering is arbitrary to begin with. Then renumber the x's accordingly. For all $1 \le J \le k - 1$ and all α we suppose that

$$\Delta_J^\alpha q_k = \sum_{i=1}^{k-1} a_{ki} \Delta_J^\alpha q_i$$

Note first that we can always write

$$q_k = \sum_{i=1}^{k-1} A_{ki} q_i + \sum_{j=k+1}^{m/2} B_{kj} q_j + u_k$$

where q_j, $k+1 \leq j \leq m/2$ is not caused by q_i, $1 \leq i \leq k-1$, with A_{ki} possibly 0. For consider any causal equation of this form where some of the q_j are caused by some of the q_i. To find a true causal law of the required form simply substitute for each of the unwanted q_j an expansion in a set of causes of q_j, all of which occur prior to all q_i. From this it follows from our lemma that for all J such that $i \leq J \leq k-1$

$$\Delta_J^\alpha q_k = \sum_{i=1}^{k-1} A_{ki} \Delta_J^\alpha q_i$$

We need to show that $A_{ki} = a_{ki}$. Consider first $\Delta_L^\alpha q_k$, where $1 \leq L \leq k-1$ and q_L is causally posterior to all other q_i for $1 \leq i \leq k-1$:

$$\Delta_L^\alpha q_k = a_L \Delta_L^\alpha q_L = A_{kL} \Delta_L^\alpha q_L$$

where the first equality comes from the assumption that the equation for q_k predicts first differences correctly and the second from the true law for q_k. It follows that $a_{kL} = A_{kL}$.

Next consider $\Delta_{L'}^\alpha q_k$, where $i \leq L' \leq k-1$ and q_L' is causally posterior to all other q_i for $l \leq i \leq k-1$ except for L:

$$\Delta_{L'}^\alpha q_k = a_{kL} \Delta_{L'}^\alpha q_L + a_{kL'} \Delta_{L'}^\alpha q_{L'} = A_{kL} \Delta_{L'}^\alpha q_L + A_{kL'} \Delta_{L'}^\alpha q_{L'}$$

for the same reasons as before. Since $a_{kL} = A_{kL}$, it follows that $a_{kL'} = A_{kL'}$: And so on for each coefficient in turn. QED.

Notice, however, that this theorem is not very helpful because it will be hard to tell whether an equation has indeed predicted first differences rightly. That is because we will not know what $\Delta_J^\alpha q_j$ should be unless we know how variations in u_J affect q_j and to know that we will have to know the causal relations between q_J and q_j. So in order to judge whether each of the q_j affects q_k in the way hypothesized, we will have to know already how they affect each other. If we happen to know that none of them affect the others at all, we will be in a better situation, since the following can be trivially derived from theorem 2a.

Theorem 2b A regression equation $x_k \overset{c}{=} \sum_{j=1}^{k-1} a_{kj} x_j + \Psi_k$ in which no right-hand side variable causes any other is causally correct iff for all α and J, $\Delta_J^\alpha q_k = a_{kJ} \Delta_J^\alpha u_J$.

We can also do somewhat better if we have a complete set of hypotheses about the right-hand-side variables. To explain this, let me define a complete causal structure that represents an ECLDSwPM, $\langle Q = \{q_1, \ldots, q_{m/2}, u_1, \ldots, u_{m/2}\}, \mathrm{CL}\rangle$ as a pair $\langle X = \{x_1, \ldots, x_n : 1 \leq n \leq m/2\}, \mu, \mathrm{CLH}\rangle$, where μ is a probability measure over the x's and where the causal law hypotheses, CLH, have the following form:

$$x_1 \overset{c}{=} \Psi_1$$
$$x_2 \overset{c}{=} a_{21}x_1 + \Psi_2$$
.
.
.
$$x_n \overset{c}{=} \Sigma_{j=1}^{n-1} a_{nj}x_j + \Psi_n$$

where $\Psi_j \perp x_k$, for all $k < j$. In general $n < m/2$. Since the ordering of the q's has no significance, we will again suppose that they are ordered so that q_j is represented by x_j. Now I can formulate the following.

Theorem 2c   If for all x_k in a complete causal structure, $\Delta_J^\alpha q_k = \Delta_J^\alpha x_k$ as predicted by the causal structure for all α and J, $1 \leq J \leq n$, then all the hypotheses of the structure are correct.

For the proof we need some notation and a convention. What does the causal structure predict about differences in q_k for $\Delta_k^\alpha u_k$? I take it to predict that $\Delta_k^\alpha q_k = \Delta_k^\alpha u_k = \alpha$. (To denote a predicted difference I use Δ', with Δ reserved for real differences (i.e. those that follow from the causal system being modelled in the causal structure). So the antecedent of theorem 2c thus requires that for all J, $1 \leq J \leq n$, $\Delta_J^\alpha q_k = \Delta_J'^\alpha x_k$.

Proof   Consider the kth equation in the structure:

$$x_k \overset{c}{=} \sum_{i=1}^{k-1} a_{ki}x_i + \Psi_k$$

We need to show that

$$q_k \overset{c}{=} \sum_{i=1}^{k-1} a_{ki}q_i + \sum_{j=k+1}^{m/2} b_{kj}q_j + u_k$$

where q_i does not cause q_j for $1 \leq i \leq k-1$ and $k+1 \leq j \leq m/2$. We know that for some $\{A_{ki}\}, \{B_{ki}\}$

$$q_k \overset{c}{=} \sum_{i=1}^{k-1} A_{ki}q_i + \sum_{j=k+1}^{m/2} B_{kj}q_j + u_k$$

where q_i does not cause q_j for $1 \leq i \leq k-1$ and for j such that $k+1 \leq j \leq m/2$ and $B_{kj} \neq 0$. So we need to establish that there is a set of A_{ki}

such that $A_{ki} = a_{ki}$ for all i such that $1 \leq i \leq k - 1$. We do so by backwards induction: show first that the coefficient of x_{k-1} is correct and work backwards from there. Note for the proof that since q_i, $1 \leq i \leq k - 1$, does not cause q_j, for any j such that $k - 1 \leq j \leq m/2$ and $B_{kj} \neq 0$, $\Delta_i^{\alpha} \sum_{j=k+1}^{m/2} B_{kj}q_j = 0$ for $l \leq i \leq k - 1$.

Inductive base to show $A_{kk-1} = a_{kk-1}$

$$\Delta_{k-1}^{\alpha} q_k = \sum_{i=1}^{k-1} A_{ki} \Delta_{k-1}^{\alpha} q_i = \sum_{i=1}^{k-1} A_{ki} \Delta'^{\alpha}_{k-1} q_i = A_{kk-1}\alpha$$

$$= \Delta'^{\alpha}_{k-1} q_k = \sum_{i=1}^{k-1} a_{ki} \Delta'^{\alpha}_{k-1} q_i = a_{ki}\alpha$$

So $A_{kk-1} = a_{kk-1}$.

Inductive argument: given $A_{k,p+s} = a_{k,p+s}$ for $1 \leq s < k - 1 - p$, to show $A_{kp} = a_{kp}$, consider what happens given Δ_p^{α}. Using the reduced form for q_i plus the assumption that all first difference predictions are right, and the fact that $\Delta'^{\alpha}_p q_i = 0$ for $i < p$, we have:

$$\Delta_p^{\alpha} q_k = \sum_{i=1}^{k-1} A_{ki} \Delta_p^{\alpha} q_i = \sum_{i=1}^{k-1} A_{ki} \Delta'^{\alpha}_p q_i = \sum_{i=p}^{k-1} A_{ki} \Delta'^{\alpha}_p q_i$$

$$= \sum_{i=p}^{k-1} A_{ki} \Delta'^{\alpha}_p u_p \sum_{l=p}^{i-1} a_{il} \sum_{m=p}^{l-1} a_{lm} \cdots$$

$$= A_{kp}\alpha + \sum_{i=p+1}^{k-1} A_{ki} \alpha \sum_{l=p}^{i-1} a_{il} \sum_{m=p}^{l-1} a_{lm} \cdots$$

$$= \Delta'^{\alpha}_p q_k = \sum_{i=p}^{k-1} a_{ki} \Delta'^{\alpha}_p q_i = a_{kp}\alpha + \sum_{i=p+1}^{k-1} a_{ki} \alpha \sum_{l=p}^{i-1} a_{il} \sum_{m=p}^{l-1} a_{lm} \cdots$$

By hypotheses of the induction $A_{ki} = a_{ki}$, for $p + 1 \leq i \leq k - 1$. Hence $A_{kp} = a_{kp}$.

There is one important point about exogenous variables that we need to be clear about to understand the significance of the theorems. By definition, $\Delta_j^{\alpha} q$ is the difference in q given a difference in u_j with all other exogenous quantities in the system, not just those in the structure, held fixed. It is easy to see why. (Notice that this echoes the concerns expressed in the cautionary lessons at the end of ch. 13 of this book.) Consider a six-quantity system:

$$q_3 \overset{=}{c} u_3$$
$$q_1 \overset{=}{c} a_{13}q_3 + u_1$$
$$q_2 \overset{=}{c} a_{23}q_3 + u_2$$

and a two-variable causal structure to represent it

$$x_1 \; c^= \; \Psi_1$$
$$x_2 \; c^= \; c_{21}x_1 + \Psi_2$$

These will be true viewed just as regression equations given

$$c_{21} = a_{23}a_{13}/\left[1 + a_{13}{}^2\right]$$

and

$$\Psi_2 = (a_{23}/a_{13} - a_{23}a_{13}/\left[1 + a_{13}^2\right])q_1 + u_2 - (a_{23}/a_{13})u_1.^{14}$$

If u_1 varies while u_2 and u_3 do not, then we will see rightly that the equation for x_2 is not correct. But if as u_1 varies, u_3 varies as well in such a way that $a_{23}\Delta u_3 = c_{21}\Delta u_1$, then the equation for x_2 will produce the right first difference predictions for x_2. That is why, to get a proper test for the equation, we must consider variation in exogenous variables in the structure while all other exogenous quantities in the system and (also in the structure) remain constant.

This makes the results more difficult to put to use than we might have hoped. In the first place, for the theorems to apply at all, we need to know that we are dealing with an epistemically convenient system – one for which the exogenous factors have no cross-restraints. But it is hard enough to know about the cross-restraints on the exogenous causes for a set of putative causes we are considering in our structure, let alone for a lot of possible causes in the system that we have no idea of.

Suppose though that we do have good reason to think that the system we are studying is epistemically convenient (or we are prepared to bet on it). How would we use the theorems to which that entitles us? The most straightforward application of the theorems to test a hypothesis about the causes of q would consider variations in the exogenous factors for q's putative causes holding fixed all other exogenous factors, where these have to include all other exogenous factors in the system. So we would have to know what these factors are. Again, it is hard enough to know what the exogenous causes are for factors we can identify without having to know what they are for factors we do not know about.[15]

I take it that this is the chief motivation for stressing manipulation. It seems that if we vary the putative causes at will or arbitrarily the variation will not match any natural variation in other exogenous factors. But we know that is

---

[14] Recall that for $x_2 = c_{21}x_1 + \Psi_2$ to be a regression equation, $\langle x_1, \Psi_2 \rangle = 0$. I assume here that the u's have mean 0, variance 1 and $\langle u_i, u_j \rangle = 0, i \neq j$.

[15] As we know, randomized treatment-control experiments are designed to allow us to get around our lack of knowledge of the exogenous factors for missing factors. But the knowledge that we have succeeded in the aims of randomizing even when we have used our best methods is again hard to come by.

not true. Coincidences happen, even when the variation is chosen completely arbitrarily – which we know at any rate is hard to achieve due to placebo effects, experimenter bias and the like. For these theorems, exactly what is required is the right kind of variation, no more and no less. So the emphasis on manipulation for invariance tests of causality is misplaced, except as a not-100 per cent-reliable methodological tool.[16]

10.4   Final remark

We are interested in whether invariance (or some substitute) under intervention is a sure sign of correctness in a causal claim. I have formalized two distinct senses commonly in use for each of the three concepts involved. That means there are eight versions of the question using just the concepts defined here. I have answered the question for only three: (1) for invariance under causal-law variation and correctness *simpliciter*, the answer (with caveats) is 'yes'; (2) for invariance under intervention/variation of values and correctness *simpliciter*, the answer is 'no'; and (3) for prediction of first differences under intervention/variation of values, the answer for prediction of first differences is 'yes'.

We can clearly carry on pursuing the other combinations, or devise modifications of the concepts that might serve better in hunting good tests. With respect to the concepts deployed here, one in particular is fairly central: that is the version of the question involving parameter correctness under first difference prediction. That is because of our usual epistemic situation. First, when a hypothesis does not involve a full set of determining factors, we are forced to look at the predictions about first differences since it makes no sense to ask whether the hypothesis is invariant or not; and correlatively, we can demand only correctness in the parameters on offer, not full correctness. Second, when the system under study is not epistemologically convenient, we are forced to use causal-law intervention to get the variation we need. I take it the answer for this particular combination is 'yes' – with caveats. But, as with any answer, we need a clear statement of the caveats and a convincing proof.

There is a division among philosophers of science between those who believe that formalization is essential to understanding and those who do not. Here I have been arguing on the side of the formalizers. The point for me of studying the relations between causality and invariance is to make better causal judgements; and if different ways of making our theses precise matter to how we make our judgements, then we had better be precise. We have seen that they do matter. Invariance under intervention is a fine test for causality if the intervention involves looking at what happens in different causal systems, but not if it involves looking at different situations governed by the same system

---

[16] As, of course is widely recognized in the experimental literature in the social sciences.

of laws. Or, when we do look at different situations, what counts as a test of a causal hypothesis when none of the putative causes causes any of the others will not serve when some do cause others.

Formalization is, however, nowhere near the end of the road. We still face the traditional problem of what all these precisely defined concepts mean in full empirical reality. In particular what is the difference between a variation in the value of a putative cause that arises from a variation in the causal system governing it versus one that arises from a variation in an exogenous cause that operates within the original system? Imagine I am about to do a randomized treatment-control experiment. How do I judge whether my proposed method of inducing the treatment fits one description or the other? I do not know how to answer the question. Perhaps indeed the distinction, which makes such clear sense conceptually, does not fit on to the empirical world it is intended to help us with. Formalization is, to my mind, the easy (though necessary) part of the job. Our next task is to provide an account of the connection between our formal concepts and what we can do in practice.

Part III

Causal theories in economics

11   Preamble

The first four chapters in part III focus on hunting causes in economics; the last on using them. None, I am afraid, even starts on the job I urge to be so important: showing how our methods for hunting causes can combine with other kinds of knowledge to warrant the uses to which we want to put our causal claims.

I look at two different methods employed in economics for hunting causes. The first uses econometric techniques, the second, theoretical models. The econometric techniques discussed are a caricature of what happens in economics. I look at only one author who claims to get causes from probabilities – Herbert Simon – and then only at the simplest imaginable cases. That is because I think Simon has a clear idea of the difference between selecting a model that is accurate about the underlying probability from which the data are drawn versus one that is accurate in describing causal relations. I want to focus on this difference without intertwining it with any of the pressing problems that must be treated in trying to ensure correct estimation of probabilities.

The two chapters on Simon here might be seen as part of a trilogy: Simon as interpreted by me, where I suppose that the trick in getting causes from probabilities depends on the characteristics of the 'exogenous' variables; Simon as interpreted by Kevin Hoover,[1] who supposes that what matters are the parameters that describe the quantities we can directly control; and Simon as interpreted by Damien Fennell, who takes the idea of separate mechanisms to be central.[2] Fennell's paper grew out of LSE's Arts and Humanities Research Board-sponsored project on Causality: Metaphysics and Methods, as did many of my own papers. Fennell's Simon-inspired account of how to get causes from probabilities can be usefully compared with the modularity views discussed earlier in part II.

Another paper from the same project by Julian Reiss[3] also studies econometric methods for hunting causes, in this case by using instrumental variables. Reiss follows the strategy I advocate in ch. 5 proving a kind of 'representation theorem'. We came to see the importance of this kind of representation theorem

---

[1] That is, Simon as interpreted by Hoover as Hoover is interpreted by me.
[2] Fennell (2005a).   [3] Reiss (forthcoming(a)) and Reiss (forthcoming (b)).

during our work on the project. Reiss first provides a formal characterization of the kinds of causal systems that can be treated using an instrumental-variables method; second, he describes formally what the method consists in; then he proves that the method delivers correct results if ideally implemented.

It should be noted that both the Cartwright and the Fennell version of Simon-inspired methods as well as Reiss's account of the instrumental variables approach are what in ch. 3 are called 'narrow-clinching' methods. They are highly restricted in scope and have very demanding requirements for their successful implementation. Still, to the extent that they can be carried out successfully, the conclusions follow deductively. This bears out my repeated claim that valid causal inference is difficult, but it is not impossible. With respect to the difficulties I would especially like to direct attention to the closing section of ch. 13, 'Cautionary lessons for econometrics'. There I point out that valid causal inference depends on a great deal of causal input – 'maintained assumptions' or antecedent causal knowledge – and, in keeping with the recognition from part I that causal relations come in a variety of kinds, this must even include assumptions about exactly what kind of causal relation is at stake. This makes causal inference and warrant for use all the more difficult. But there is no use shying away from it. Policies will be taken and whether they work will depend on what causes are operating and how. So somebody will at least implicitly make a great many causal assumptions, even if the econometrician will not. My own hope is that the formulation and warranting of these assumptions will be done to the highest scientific standards that are practicable and that econometricians will be part of this enterprise.

Chapter 15, 'The vanity of rigour in economics: theoretical models and Galilean experiments' takes up an old idea of mine[4] endorsed by Daniel Hausman, Uskali Mäki and others,[5] in defence of unrealistic models in economics. It argues that the defence often misfires. The original idea is that many idealized models in economics can be seen as thought experiments structured like Galilean experiments. They aim to isolate a single cause to learn what it does when acting 'on its own'. Here I worry that the results of these models are generally overconstrained. They depend not just on idealizations that isolate the factor under study but also on unrealistic assumptions necessary to provide enough structure to draw conclusions deductively. And overconstrained conclusions – ones that are too narrow – can be very misleading for policy.[6]

The final chapter is on using causes. When we deliberate about policy we should like to estimate as best possible what would happen if this or that policy were implemented in this or that way. As often noted earlier in this book,

---

[4] Cartwright (1989).   [5] See Hausman (1992) and Mäki (1992).

[6] For more on these kinds of thought experiments in economics see Reiss (forthcoming(b)). See also Sugden (2000). For objections to taking even the models that look like Galilean experiments as ways of warranting tendency claims, see Alexandrova (2005).

there is a long tradition of associating causality with manipulation and effective strategy. Chapter 16 points out that the kinds of counterfactuals that are central in hunting causes are not the ones we need to know about for using them and there is little explicit theory that bridges between them. The chapter looks in particular at work on causality by economists James Heckman, David Hendry, Kevin Hoover and Stephan LeRoy. It is a companion to a joint paper by Julian Reiss and me not contained in this volume.[7]

[7] See Reiss and Cartwright (2003).

12   Probabilities and experiments

12.1   Introduction

How do we test econometric models? The question invites a series of lessons and precautions about statistical inference. Before we take up these lessons we need to answer a prior question. What kind of information is the econometric model at hand supposed to represent? I want to focus on two broad answers: (1) the econometric model summarizes information about a probability distribution, and in addition (2) the model makes claims about causal relations. The second project clearly brings with it new and difficult problems. Even if we had full information about the probability distribution over a set of variables, that would not tell us the causal relations among them. Nevertheless probabilities may be a useful tool for inferring causal structure even if they cannot do the job on their own.

It is widely acknowledged that probabilities are most useful as a tool for causal inference in the context of a controlled experiment. For many, information about what would happen in an ideal controlled experiment is enough: we can count that as just the information we are looking for under the heading 'causation'. Suppose we take that point of view. That still leaves us a long way from conventional econometric models, which describe statistical relations in population data and not in data generated in the highly controlled environment of an experiment. Or does it? In her book *The History of Econometric Ideas* Mary Morgan[1] claims that the early econometricians were excited and optimistic about the discipline they were developing because they believed that their new statistical techniques provided 'a substitute for the experimental method' (p. 96). I want to explain how that can be the case: how, if the circumstances are felicitous, the probabilities that are represented in conventional econometric models will track those that would arise in a controlled

This paper first appeared in *Journal of Econometrics*, 1992. An earlier version was presented at the conference on 'The Significance of Testing' at Tilburg University, 1991. Thanks to Jordi Cat, Daniel Hausman and James Woodward for their assistance and the LSE Research Fund and the Centre for the Philosophy of Natural and Social Science for support.
[1] Morgan (1990).

experiment. That is, I want to show how we can infer causal information from econometric data, not only facts about the strengths of causal influence but also facts about whether a causal relation exists at all and if so, in what direction it operates.

I say 'infer' because I have in mind a very strong relation, sometimes called 'bootstrapping'.[2] Received wisdom teaches that the standard method for testing throughout both the natural and the social sciences is the hypothetico-deductive method: the hypothesis under test couples with other background information to imply claims about the data. If the predictions fail, the hypothesis is suspect. If the predictions obtain, we do not know what to conclude since a variety of incompatible hypotheses will imply the predictions equally well; but we are inclined to think that the hypothesis is confirmed to some degree. Bootstrapping is just the converse of this process. In a bootstrap argument the data couple with the background information to imply the hypothesis. This is scientific testing at its best . . . and obviously it is difficult to achieve. It is surprising that the founders of econometrics hoped for such an unusually powerful tool. (Bootstrapping methods are all cases of what in ch. 3 are called 'narrow-clinching' methods.)

The point I want to argue here is that the econometricians' hopes are in principle borne out. Occasionally – very occasionally – the conventional methods that we use to identify the parameters of econometric models can be used to test – in this very strong sense of test – the kinds of causal claims we would like to see represented in these models. I have made this kind of argument elsewhere, but I think I can make the point more simply here. The ideas are fundamentally those of Herbert Simon.[3] I will look at two developments of them, first by T. J. Cooley and S. F. LeRoy and the second by Kevin Hoover. The well-known paper by Robert Engle, David Hendry and Jean-François Richard on the stability of econometric laws will play a part as well.[4] The key to the argument is the controlled experiment.

12.2   Causes and probabilities

A large part of econometrics today was developed at the Cowles Commission in Chicago during and after World War II. Workers at the Cowles Commission supposed that it is possible to learn about causes from statistics, but that one needed to add in a lot of theory in order to do so. Nowadays the US school advocating vector-autoregression techniques (using the work of Clive Granger, Christopher Sims and others) tries to establish causal relations just from statistics, with no prior inputs. Cooley and LeRoy attack this attempt to carry on macroeconomics atheoretically. Their aim is to defend the importance

---

[2] Glymour (1980).   [3] Cartwright (1989); Simon (1957), ch. 2.
[4] Cooley and LeRoy (1985); Hoover (1990; 1991); Engle, Hendry and Richard (1983).

of a standard Cowles Commission assumption about exogenous variables. As Cooley and LeRoy characterize it: 'An exogenous variable is one determined outside the model, while an endogenous variable is one determined inside the model.'[5] The exogenous variables can be causes of the endogenous ones, but not the reverse. The assumption that Cooley and LeRoy take to be crucial is the familiar supposition that 'all the exogenous variables are uncorrelated'. This is of course a substantial empirical assumption and may in many – indeed most – cases be unwarranted. But we are at great advantage when the assumption does hold, for it contributes to three distinct projects:

(i) *Statistical inference*: lack of correlation among the exogenous variables is one of the key conditions that ensures that various methods of statistical inference have desirable characteristics.

(ii) *Identifiability*: uncorrelatedness of exogenous variables is one of a set of conditions commonly supposed at the Cowles Commission (like linearity and time ordering) that guarantees that the parameters of an econometric model can be identified from the appropriate probabilistic information.

(iii) *Causal inference*: as I have indicated, it was commonly supposed in the early Cowles work that econometrics teaches us about causal relations. The coefficients of an econometric model are supposed to represent the strength of effect of the independent variables on the dependent variables. The independent variable is a cause of the dependent variable just in case the corresponding parameter is not zero. Uncorrelatedness of the exogenous variables is crucial here too.

The first two claims are uncontroversial. The third is not. The question is: why can we assume that we can read off causes, including causal order, from the parameters in equations whose exogenous variables are uncorrelated? After all, equations are just expressions of complicated forms of association and it is widely acknowledged that mere association is not causation. Cooley and LeRoy's answer involves hypothetical experiments. For ease of comparison, I will convert their simple two-variable examples to the notation that Hoover uses, involving price and money; also I will renumber the equations. Otherwise, I quote directly:

Suppose we start with the model

$$M = v, \tag{1}$$

$$P = aM + \xi + d, \tag{2}$$

where v and ξ are correlated (contrary to the Cowles Commission requirement).

If the analyst were willing to assume that the correlation between v and ξ occurred because v determines a component of ξ – i.e., $\xi = \lambda v + \mu$ – then (1) and (2) could be

---

[5] Cooley and LeRoy (1985), p. 291.

written as

$$M = v, \tag{3}$$
$$P = aM + \lambda v + \mu + d, \tag{4}$$

with μ and v uncorrelated. Since M is exogenous in this set-up, the effect of a change in M on P is well-defined: $dP = (a + \lambda)dv$. Here [the size of this effect] $a + \lambda$ could be estimated by regression.

If on the other hand, the analyst were willing to specify that the correlation between ξ and v of (1)–(2) owes to a causal link in the reverse direction, the system could be rewritten

$$M = \varphi + \delta\xi, \tag{5}$$
$$P = aM + \xi + d, \tag{6}$$

with ξ and φ uncorrelated. Now the question 'What is the effect of M on P?' is not well-posed, since the answer depends on whether the assumed shift on M is due to an underlying change in φ (in which case the answer is: $dP = ad\varphi$) or in ξ (in which case the answer is: $dP = (a\delta + 1)d\xi$).[6]

Following this line of thought, and also their similar way of talking throughout, it might appear that Cooley and LeRoy offer an excessively operationalist proposal:

(1) If X causes Y, there must be an unambiguous answer to the question: 'How much does Y change given a unit change in X?' (2) But if the exogenous variables are not uncorrelated, we cannot find an unambiguous answer. (3) Hence X causes Y (if and) only if the equations governing X and Y (satisfy all the other Cowles requirements for causal structures and) have uncorrelated errors.

I think this is a mistaken reading. Recall their and my remarks about experiments. Here is what Cooley and LeRoy say (p. 295):

Because exogenous variables are conceived to be determined outside the model, hypothetical experiments consist of varying one exogenous variable *cet. par.*, and determining the effect on the endogenous variable. In order that such experiments be well defined it is essential to specify precisely what is invariant under the hypothesized intervention; this is the role of the uncorrelatedness assumption.

Think how an ideal experiment for the hypothesis M causes P should work. In an experiment we vary M and we look to see if P varies correspondingly. In the most basic case for the results to be decisive we must be assured that M can be changed without in any other way changing P. Look back at equations (1)–(2). Can we use v as a control variable for M in an experiment to test M's effect on P? That depends on whether changes in v can bring about changes in ξ or not. In the case where $v = \varphi + \delta\xi$, as in equations (5)–(6) we have no

---

[6] Cooley and LeRoy (1985), pp. 291–2.

assurances that the change in v is not produced in a way that produces a change in ξ, so v is not a good control variable. But if equations (1)–(2) are not true, but rather ones just like them but with v and ξ uncorrelated, v is a candidate for a control variable. And if equation (2) is a correct expression of the associations in nature, we can see from the equation what will result in our experiment. If $a \neq 0$, then a variation in the value of M will indeed be followed by a variation in that of P, and not so if $a = 0$. But if ξ and v are uncorrelated, we can identify a from the probabilities. So if we knew that the exogenous variables for the two-variable P–M model were uncorrelated, we could infer whether M causes P by looking at population data.

Although Cooley and LeRoy stress the importance of the lack of correlation between the exogenous variables in this argument, we should not forget that lack of correlation is not a sufficient condition even in a two-variable model. This is easily illustrated by the familiar observation that equations (1) and (2) can be rewritten in a way that suggests exactly the opposite causal ordering, with the exogenous variables in the new equations still uncorrelated. That is, equations (1) and (2) (with $\langle v \rangle = \langle \xi \rangle = 0$ and $\langle \xi v \rangle = 0$) are equivalent to

$$P = \delta \tag{1'}$$
$$M = bP + \omega + k \tag{2'}$$

where

$$\delta = av + \xi + d$$
$$\omega = -\xi/a - k - d/a$$
$$k = -\langle \xi^2 \rangle/ad - d/a$$
$$b = 1/a$$
$$\langle \delta \omega \rangle = 0$$

The equivalence of (1) and (2) with (1') and (2') does not, however, undermine the possibility of drawing conclusions about the causal order of M and P given the right background information. In order to read the experimental results from equations (1) and (2) we must not only know that ξ and v are uncorrelated, but we must also know that (i) v causes M, (ii) ξ represents all causes of P other than those that operate through M, and (iii) neither v nor ξ causes the other nor do they have causes in common. But if these conditions are satisfied for v and ξ in equations (1) and (2) they cannot at the same time be satisfied for δ and ω of equations (1') and (2'). (I sketch a proof in the footnote.)[7]

---

[7] We suppose that causal structures can be represented by directed acyclic graphs. Each event is a linear function of a set of variables, one from each of the paths leading into the node that represents it, with the usual relations among path coefficients. Equations of this kind will be called 'causal equations'. It is assumed that all valid functional relations are generated from causal equations. For a two-variable model we consider the entire graph representing the causes and effects of both variables. Assuming that equations (1) and (2) are generated by causal equations from a two-

To summarize, Cooley and LeRoy talk about hypothetical experiments. The experiments they consider are merely imaginary. How are Cooley and LeRoy to tell what will happen in a purely imaginary experiment? They will look at their equations. But the equations will tell what would happen in a proper experiment only if the exogenous variables in the equations are uncorrelated. I mentioned that the Cowles Commission insisted on other restrictions as well before a system of equations could be considered to reflect a genuine ordering. I will not lay these out here.[8] Let us just call any system of equations that satisfies all these constraints and has uncorrelated errors, a system of canonical form. As I see it then, the Cooley and LeRoy proposal is this:

If a system of equations is in canonical form, we can tell by looking at the equations what would happen in an ideal experiment. Since equations of canonical form can be identified from the probabilities which can in turn be reliably inferred, we can thus infer causes from population statistics.

But clearly the statistical relations between the variables in the model will not tell whether the variables determined outside the system are correlated or

variable model in P and M, the conditions for reading experimental results in the way suggested from these equations are the following: (i) v causes M; (ii) ξ causes P; (iii) no paths enter P except via ξ or M; (iv) v does not cause ξ, ξ does not cause v, and there is no φ that causes both v and ξ. These conditions cannot be simultaneously satisfied for v and ξ of equations (1) and (2) and for δ and ω of equations (1') and (2') consistent with the requirement that $\delta = av + \xi + d$.

To see this consider the sets of causal equations for (1) and (2) under the assumption that these conditions are satisfied in both cases:

(i) $v = \sum a_i^0 x_i^0 + k_1$

(ii) $\xi = \sum b_i^0 y_i^0 + \lambda \delta + k_2$

(iii) $\delta = \sum c_i^0 w_i^0 + k_3$

(iv) $\left\{ x_i^{n-1} = \sum a_{ij}^n x_{ij}^n + k_4^{in-1} \right\}$

(v) $\left\{ y_i^{n-1} = \sum b_{ij}^n y_{ij}^n + k_5^{in-1} \right\}$

(vi) $\left\{ w_i^{n-1} = \sum c_{ij}^n w_{ij}^n + k_6^{in-1} \right\}$

(vii) $\left\{ z_p = \sum f_{pmi} x_i^m + g_{pkn} y_n^k + h_{pqr} w_r q + d_p v + \gamma_p \delta + k_7 p \right\}$

where the z_p's are any variables that are not causes of ξ or v or δ. Constraints (i) to (iv) for both sets of equations require (assuming $\lambda \neq 0$ which is the more difficult case) for all $i, j, k, l, p : x_i^j \neq y_k^l, x_i^j \neq w_k^l, x_i^j \neq \delta, x_i^j \neq \xi, y_i^j \neq v, w_i^j \neq v, z_p \neq x_i^j, z_p \neq y_i^j, z_p \neq w_i^j, z_p \neq \delta$. Consider how $a\delta = v + \xi + d$ might arise. No subset of (i)–(vii) that includes any element from the set (vii) will produce an equation in δ, v, and ξ alone, since z_p for the largest p can never be eliminated. Therefore we must use a set of equations from (i)–(vii) containing both (1) and (2). But inspection of the associated matrix readily shows that (assuming v is not a constant) no set including (i) will produce an equation in δ, v, ξ alone since some x_i^j will always occur.

[8] In *Nature's Capacities and Their Measurement* (1989) I spell out a full set of conditions which are jointly sufficient for the inference of causal structure from a set of equations. See also ch. 13 in this book.

not. Nor will these statistics tell us the additional information that we will need about the causal roles of M and P. For Cooley and LeRoy that is why we need theory.

Turn now to Simon himself. Simon presents a method for reading causal order from linear equations. For equations (1)–(2) his method determines the order M causes P; for equations (1')–(2') just the reverse: P causes M. Yet the two sets of equations are observationally equivalent in that they represent the same solutions in P and M. Nevertheless, Simon maintains, there is a difference between them. If a parameter indicated for the M process (i.e. v) were to change without any alterations in the parameters indicated for the P process (i.e. a and ξ), M would change and P must as well. But if a parameter of the P process changes while those for the M process remain fixed, P would change but M would not. (Presumably the possibility of the parameters indicated for the P process changing while those indicated for the M process remain fixed and the converse is taken to be likely given the lack of correlation between v and ξ.) Simon argues that this difference between the two equation sets represents an objective, operationalizable difference between situations correctly represented by them. Which situation is the real situation can be determined by an experiment which controls 'directly' either v or ξ and then observes the indirect changes that follow in P and M.

At this point there are two alternative ways to proceed, depending on how we interpret my expression 'the parameters indicated' for the P and M processes. The weak interpretation is that they are the parameters used in a given set of equations to represent the marginal and conditional distributions in P and M. In that case Simon's hypothetical experiments need not match the controlled experiments that I have suggested we take as a standard for causal testing. That is because the procedures for these experiments require that it is not just arbitrary parameters describing the P and M distribution that change independently of each other, but rather those describing the causes of P and M. So there is no ground for taking the order dictated by such an experiment to be a causal order, or to have any significance at all. Moreover, unless we assume, for example, that v represents a cause of M, it is not clear why we should think that it is possible to control it directly and thereby control M indirectly as Simon requires.

Alternatively, we could assume that Simon is taking this additional causal information for granted. (For my own account of Simon, see ch. 13 in this book.) As I understand it this is what Kevin Hoover wishes to do in his development of Simon's work. For Hoover, in urging that we can draw conclusions about causal order from observations on interventions that change the probability distribution, argues that we can do so '*crucially* [his italics] if we can attribute the interventions either to the money determination process or to the price determination process'.[9] If we do read Simon in this way, then a proof of the

9 Simon (1991), p. 384.

kind I described in footnote 1 would complete his argument that the two equation sets represent operationizably different situations.

12.3   The stability of causal order

Having defended the possibility of using population statistics to test for causal order, I now want to urge a caution. For most of our purposes knowing a set of causal relations is not enough; we need to know which of these causal relations is stable. Perhaps in the best of circumstances we can get from population probabilities to experimental results. But are experimental results, even in a totally ideal, perfectly controlled experiment, enough to secure what we need? The point of making causal claims is to provide a guide for what the putative cause can achieve not in the narrowly confined conditions of the controlled experiment but rather in the mixed and various conditions we encounter outside the test situation. We have still to confront the traditional problem: what licenses the move from the results we see in experimental settings to applications we want to make in ordinary non-experimental settings? The problem is nicely illustrated by Engle, Hendry and Richard.[10]

Once the distributions for the exogenous variables are provided, equations (1) and (2) determine the joint probability distribution in P and M. I assume for simplicity that it is normal, with $v \sim N(0, \sigma_v^2)$, $\xi \sim N(0, \sigma_\xi^2)$, and $\langle \xi, v \rangle = 0$. The marginal distributions are then given by

$$D(M) \sim N\left(0, \sigma_v^2\right) \tag{7}$$
$$D(P) \sim N\left(d, a^2\sigma_v^2 + \sigma_\xi^2\right) \tag{8}$$

This suggests just the result we might expect having established the causal relation M causes P, that by changing $D(M)$ (by changing σ_v^2) we will be able to produce desired changes in $D(P)$. But, following Engle et al. consider an alternative expression for the bivariate normal distribution in P and M in terms of its covariance matrix, $\Omega = \{\omega_{ij}\}$, where

$$\sigma_v^2 = \omega_{mm} \tag{9}$$
$$\sigma_p^2 = \omega_{pp} \tag{10}$$
$$\sigma_\xi^2 = \omega_{pp} - \left(\omega_{pm}^2/\omega_{mm}\right) \tag{11}$$

Consider now the proposal to change σ_v^2 while leaving σ_ξ^2 the same. What will happen to $D(P)$? As Engle, et al. point out, we do not know. Even if we assume that after the shift the new distribution is again bivariate normal, the new parameters will be unsettled. Nature might, for instance, be unconcerned with the ratio of ω_{pm} to ω_{mm} and arrange that with σ_ξ^2 fixed, ω_{pm} changes with ω_{mm} in just the right way to allow ω_{pp} to say fixed as well. Or not. The current

[10] Engle et al. (1983).

distribution of M and P does not tell us what new distribution will obtain when it no longer does.

Is not this Hume's familiar problem of induction?: there is no guarantee that regularities that have been observed to obtain in the past will continue to obtain in the future? The answer is 'no'. Hume imagined a worse situation, for he envisaged that past regularities might fail in the future for no reason whatsoever. The worry here is that they may change for very good reason, for we are envisaging new background circumstances substantially different from those in which the regularity has so far been observed to obtain and we are not sure to what extent the change in circumstances will matter.

The regularities we have been concerned with are not mere associations but rather causal relations and causal orderings. Engle, Hendry and Richard distinguish between exogeneity and superexogeneity: superexogenous relations are ones in which the parameters remain invariant across a range of envisaged interventions. Exogeneity is not causality, but we might nevertheless borrow their terminology and speak of causal relations and *super*causal relations. I prefer to talk instead of 'mere' causal relations versus capacities that will be stable across a range of envisaged changes. Using this vocabulary we can say about our argument from experiment that it provides a way of establishing causal relations from probabilities but it does not provide a way of drawing inferences about supercausal relations (stable capacities).

Putting it more carefully, the point is this. If the circumstances meet the conditions for a controlled experiment, we can sometimes (given appropriate antecedent causal information) infer from the probability distribution over the variables in that situation what causal relations obtain in that situation. But we cannot infer what causal relations will obtain outside that situation. Often we do export our causal conclusions from one situation to another, but that is generally because we have some prior commitment to the view that if a causal relation obtains somewhere, it obtains on account of the exercise of some (relatively) stable capacities. In that case any situation in which we actually know the causal relations is as good as any another for drawing conclusions about new situations. What is special about a controlled experiment is not that the causal relations that obtain there are bound to obtain elsewhere, but rather that the causal relations that obtain there can be inferred from the probabilities. The controlled experiment is epistemologically privileged – it is a good situation for finding out about what causal relations obtain in it; but it is not ontologically privileged – the causal relations that obtain there have no special status with respect to other situations.[11]

'But causality without invariance', one may ask, 'of what use is it?' The arguments here suppose that the controlled experiment is special: whatever it

[11] I discuss the stability of causal relations in more detail in Cartwright (1989).

is we mean by causality, the ideal controlled experiment is the best test for it. There is a considerable philosophical literature connecting causal hypotheses with counterfactual conditionals.[12] The claim that smoking causes lung cancer in a population P supports the counterfactual (or subjunctive) conditional: 'If an individual in P were to stop smoking, his or her chance of lung cancer would be decreased.' The literature stresses that the probability of a counterfactual conditional [Prob (if X were to occur, Y would obtain)] cannot in general be equated to the conditional probability [Prob (Y/X)] that describes the population P. But the assumption is that it can be identified with certain probabilities exhibited in ideal controlled experiments. If this assumption is correct, then the causal knowledge that arises from controlled experiments is of importance in decision making: it can provide predictions about what will happen if the variables identified as causes were set at higher or lower levels. But if the methods we propose to employ for doing so will at the same time change the underlying chance set-up in such a way that the overall probability distribution is altered, additional invariance assumptions will be required as well.

Before concluding I would like to take up one further point about the distinction between invariance and causality and discuss how this distinction affects another proposed test for causal hypotheses in econometrics. The test again involves the idea of a controlled experiment; it has been advocated by Kevin Hoover in his development of Simon's work and in particular it has been applied in Hoover[13] to the problem of determining the causal direction between price and money. We begin with equations (1) and (2) with $\langle \xi v \rangle = 0$. (For discussion of another of Hoover's accounts see ch. 14 in this book.) As before we assume that v causes M and that ξ represents all causes of P that do not operate through M and that ξ does not cause v nor the reverse nor do the two have a cause in common. Consider what happens when we intervene in the money determination process. Standard experimental procedure involves changing the distribution in M without intervening in any way in the processes distribution of P changes as well. Hoover adds a second test: change the distribution of M and look to see if the conditional distribution of P on M stays the same. Hoover says:[14] '[I]f *money causes price* and if the intervention was in the money-determination process, say a change in Federal Reserve Policy . . . one would expect $D(P/M)$ to be stable, although not $D(P)$.' The question is why is this a reasonable test for the hypothesis money causes price? The first test, which looks for changes in $D(P)$ on changes in $D(M)$, is warranted as a test for causality on the grounds that it tracks ideal experimental procedure. What is the justification for looking for stability in $D(P/M)$?

---

[12] Cf. Harper et al. (1981); also further discussion in ch. 16 of this book.
[13] 1991.　　[14] 1991, p. 384.

One answer is clear. If M causes P, then in a two-variable model $D(P/M)$ measures the strength of M's effect on P.[15] Clearly the question of the invariance of the strength of this influence across envisaged interventions in M is one of considerable interest in itself. But finding out the answer is not a test for causality, either in the original situation or in any of the new situations that might be created by intervention. Even if M causes P in the original situation and continues to do so across all of the changes envisaged, there is in general no reason to think that interventions that change the distribution of M will not also affect the mechanism by which M brings about P, and hence also change the strength of M's influence on P.[16]

Consider next what can be expected under the strong but not unusual assumption that the causal structure stays entirely fixed under the interventions envisaged in the money determination process. Changes in $D(v)$ are supposed to be introduced without changes in $D(\xi)$, so σ_ξ will remain the same. Under the very special conditions described, we have seen that we can take a as a measure of the causal efficacy (if any) of M on P so that it too will stay the same under the assumption that the causal structure is invariant. Trivially then $D(P/M)(\sim N(aM, \sigma_\xi^2))$ will remain the same whether or not M causes P (i.e. whether or not $a = 0$). Hence the invariance of $D(P/M)$ across a proposed change can be taken as a necessary condition for the constancy of the causal relation between M and P across that change, even though this invariance is irrelevant to determining what that relation is.

12.4   Conclusions

The suggestions of Simon, Cooley and LeRoy, and of Hoover provide a way to justify the use of correlations and regressions in naturally occurring populations as tests for causality. The argument takes the controlled experiment as the ideal test for a causal hypothesis. It shows that in some circumstances we may not need to perform the controlled experiment; nature runs it for us. The conditions are demanding, both from the objective point of view (in the two-variable model v and r must be uncorrelated) and subjectively (we must know of the uncorrelated terms that they represent independent causes of M and P respectively), so we are not often in a situation to make use of the argument. That should not be surprising. Confirming scientific hypotheses is difficult, especially when we want to employ the powerful method of bootstrapping rather than the weak kinds of tests that the hypothetico-deductive method can supply. We could not expect causal hypotheses to be any easier than others to test.

[15] In larger models it is the conditional distribution of P or M with other causes of P held fixed that would be at issue.

[16] See Cartwright (1989) and Cartwright and Jones (1991) for a fuller discussion and chs. 7 and 8 in this book.

When it comes to application we are in a more difficult situation. Even a sure test to determine if a hypothesis holds in a given setting will not guarantee that the hypothesis holds outside that setting. This is not a problem peculiar to causal hypotheses. Nor is it a problem peculiar to economics. Whether we are testing claims about the size of a physical constant or about the functional relation between temperature and pressure in an ideal gas or about the efficacy of a new drug, we need independent reasons for thinking that the lessons of the laboratory can be carried outside its walls and for determining how far and to what extent. The reasons may be very local, as I believe David Hendry advocates. They may be sweepingly metaphysical, as we see in recent work on Bayes-nets structures and probabilities,[17] where it is assumed that all causal structures are stable. Or they may be based on more limited metaphysical assumptions, as in much of the microfoundations literature in economics, where it is commonly supposed that once we get down to the level of individual decisions, expectations and preferences, the correct models will be invariant across interventions in the macroeconomic process. Or we may use empirical tests for invariance. But with these too we will need reasons for carrying the conclusion outside the regime over which the test was conducted.

[17] E.g. Spirtes et al. (1993) and ch. 6 in this book.

13   How to get causes from probabilities: Cartwright on Simon on causation

Causal relations may not be reducible to associations but associations can certainly be used in figuring them out. This includes probabilistic associations. Econometrics studies how, and this was one of the central aims of many of its founders and early workers, especially in the Cowles Commission.[1] Here I shall review a method that works for one very basic kind of causal system, based on the work of Herbert Simon.[2] The method is deductive, as I believe econometrics always aspires to be: given the right kind of background causal information, causes are deduced from probabilities. To focus on the ideas rather than the formalism, I shall illustrate with the simplest possible case involving just two variables.[3]

As I have urged throughout this volume, methods for inferring causal relations must be matched to the kinds of causal relations we are trying to infer. So one must begin by asking 'what kind of causal relation, in what kind of causal system?' The method I shall describe here is intended to permit inference to the causal principles of a linear deterministic system as characterized in ch. 10. Begin then with, first, a set of linear deterministic equations, CCL, that are candidates for the causal principles governing a set of variables of interest, $V = X \cup U$, where $X = \{x_1, \ldots, x_n\}$ and $U = \{u_1, \ldots, u_n\}$[4] and second, a probability measure over the u_i's, $P(u_1, \ldots, u_n)$, where for convenience each u_i has

---

The material in this paper comes from undergraduate lectures in philosophy of economics that I have given on and off at LSE since 1991. I would like to thank all the students who listened, discussed, puzzled out the material and helped me to make it as clear as possible over the years.
[1] Cf. Morgan (1990).   [2] Simon (1954).
[3] The case is also exceedingly simple because it is both linear and deterministic. Determinism will turn out to be very restrictive. Although probabilities are introduced over the 'exogenous' variables, in order to ensure that we can infer that the equations are causally correct, the constraints on these variables will be so strong that they cannot represent 'unknown' quantities nor represent genuine indeterminacy.
[4] The systems considered here have the same number of u's as equations. Though this is restrictive, the analyses of the chapter can be generalized to cover cases where there are more u's than equations. To keep the discussion straightforward however, I discuss only the less general case. Note that to generalize the discussion, the assumptions about the u's (variation freedom, etc.) made here must hold of the larger set of external variables.

mean 0 ($\langle u_i \rangle = 0$) and variance 1 ($\langle u_i^2 \rangle = 1$):

CCL : $x_1 = u_1$
$\quad\quad x_2 = a_{21}x_1 + u_2$
$\quad\quad \vdots$
$\quad\quad x_n = \Sigma a_{nj}x_j + u_n$

Following Damien Fennell,[5] I will call the variables in U, 'external variables'. For the moment the distinction between the x's and the u's is purely notational. With policy implications in mind, though, it should be noted that the quantities we are interested in affecting will generally be represented by the x variables since the relations between the u's are not immediately represented in the equations. But the nature of the u's is crucial to the particular method for causal inference I describe here.

The job is to provide constraints that allow two tasks to be accomplished:
1 to identify the coefficients in the candidate system;
2 to ensure that the candidate equations are causally correct.[6]

Task 1: identification is the meat of econometrics. For our simple candidate system the task is exceedingly easy. The coefficients can be identified if $\langle u_1, \ldots, u_n \rangle = \langle u_1 \rangle \langle u_2 \rangle \ldots \langle u_n \rangle$ for all combinations. Consider as an example the simplest case of a candidate system with just two variables, say money and price. (These are the variables that Kevin Hoover focuses on in his account of Simon, which I discuss in chs. 12 and 14). So $X = \{M, P\}$ and $U = \{\alpha, \beta\}$

$$M \overset{c}{=} \alpha \tag{1}$$
$$P \overset{c}{=} aM + \beta$$

Then

$$\langle MP \rangle = a\langle \alpha^2 \rangle + \langle \alpha\beta \rangle$$

assuming $\langle \alpha \rangle = 0 = \langle \beta \rangle$ and $\langle \alpha^2 \rangle = 1 = \langle \beta^2 \rangle$. Thus if $\langle \alpha\beta \rangle = \langle \alpha \rangle \langle \beta \rangle$, then $a = \langle MP \rangle$. So $a \neq 0$ if the covariance of M and $P \neq 0$. That seems a reasonable result. If the equations of system (1). are causally correct and $a \neq 0$, then M does indeed cause P and they should be expected to co-vary.

Can we assume that the equations are causally correct? No, because of the problem of equivalence. The associations between values of P, M, α and β represented in the equation set (1) are equally represented in equation set (2):

$$P \overset{c}{=} \tau \tag{2}$$
$$M \overset{c}{=} P/a - \beta/a$$
$$(\tau = aM + \beta)$$

But equation set (2) describes the opposite causal ordering for M and P.

[5] Fennell 2005b.
[6] That is, that each equation is an equation from the linear deterministic system governing the quantities of interest.

On the other hand, in equation set (2) $\langle \tau\beta/a \rangle \neq 0$. So we might hope that if a requirement is added that the variables in U have co-variance 0, the order will be unique. That hope, however, will not succeed. For consider system (2'):

$$P \overset{=}{c} \tau \qquad\qquad (2')$$
$$M \overset{=}{c} bP + \sigma$$
$$(b = a/(a^2+1))$$
$$(\sigma = (\alpha - a\beta)/(a^2 + 1))$$

Now $\langle \tau \rangle = 0 = \langle \sigma \rangle$ and $\langle \tau\sigma \rangle = 0^7$ and $b = \langle MP \rangle / \langle M^2 \rangle$. So $b \neq 0$ if and only if $a \neq 0$, which should not be surprising because whether two factors co-vary or not is independent of the order in which they are considered. Yet $a \neq 0$ suggests that P causes M and $b \neq 0$, the reverse. So the proposed constraint does not provide a unique set of linear causal relations after all. Are there any constraints that can do so? That clearly brings us to task 2.

Task 2: the task then is to find further constraints that allow
- a unique way of writing the equations;
- that can be proven to be causally correct; and
- that retains identifiability.

My own view has always been that there is no way to do this without including causal information among the constraints: no causes in; no causes out.[8] The attempt to rely on co-variance was one attempt to use constraints that are purely associational and it fails. Another related attempt is to insist that the variables in U be variation free, that is, that they can take any of their allowed values together in any combination.[9]

Let us suppose then that for the situation under study there is a way of writing the associations between the variables of interest so that one variable can be selected from each equation to be designated an external variable and that the set of external variables is variation free. The suggestion is that if there is one way to do this, there will be only one way since any different way of writing the equations will make the new external variables functions of the old, which – maybe? – means that the new external variables cannot be variation free. This idea is mistaken.

Suppose that α and β from (1) are variation free. What about τ and σ, can they take any values in combination, say T and S? A little algebra shows that this means that $\alpha = bT + S$ and $\beta = T(1 - ab) - aS$. But that is perfectly possible

---

[7] Though note that the variances of the new external variables are not 1.

[8] Cartwright (1989), ch. 2.

[9] Range $(u_1 \ldots u_n)$ = range (u_1) x $\ldots$ x range (u_n). This may sound like the constraint introduced in ch. 10 but in fact it is weaker. Chapter 10 requires that for each causally correct equation there is a special cause peculiar to it and that these special causes are variation free. Here we have instead a set of functionally – but not necessarily causally – correct equations each with a special variable – not necessarily a cause of the left-hand-side variable – peculiar to it and the special variables are variation free. A requirement similar to that from ch. 10 will be discussed below.

since α and β can take any values in combination. So variation freedom in the external variables is not enough to guarantee uniqueness.[10]

There is another related constraint that will work, however, based on the same idea, that when the equations are rewritten the external variables get mixed up in the new equations. Notice that in systems (1) and (2) not only is there one distinct variable singled out for each equation, dubbed the 'external' variable, but also when the equation for M changes from system (1) to system (2), the new equation for M in system (2) contains an external variable that in system (1) was restricted to the equation for P. This is in general true. If the external variables are mated to specific variables of interest in one set of equations, in any way of rewriting the equations some external variable will appear in an equation for a variable that is not its mate, as the following theorem says:

Theorem 1: consider a system of equations of the form of CCL where for each x_i there is at least one variable – call it u_i – that appears in the equation for x_i and in no other equation and the u_i's are not functions of one another.[11] Then any (non-trivial) linear transformation on the equations will result in a system of equations where some u_j appears in an equation for x_i, $i \neq j$.[12]

Suppose then that the equations describing the functional relations between a set of variables can be written in the form of a linear deterministic system, as in CCL. Suppose further that in the equation for each x_i there is some factor, u_i, that is unique to that equation, appearing in no other; and suppose further that the u_i's are not functions of each other. The theorem says that if there is one way to write the functional relations in this form, there is only one way to do so that preserves the form of CCL and keeps exactly the same u_i and x_i paired in each equation. In particular this is true even if the variables are allowed to be relabelled and reordered to preserve the lower triangular form. (This means that, for instance, information about time order among the variables is either not available or not taken into account in their labelling.)

But why should we think that the unique way of writing equations in accord with these constraints ensures that the equations are causally correct? So far as I can see, there is no reason at all. What then must we further assume to justify a causal reading for the unique way of writing the equations?

---

[10] For a good more detailed discussion of this point see Fennell 2005b.

[11] For all i, $u_i \neq f(u_1, \ldots, u_{i-1}, u_{i+1}, \ldots u_n)$.

[12] This was essentially proven in Cartwright (1989), pp. 37–8. A proof also follows if one considers what would happen if the result were false, that is, if there were two mathematically equivalent but non-identical systems of equations of form CCL for the same x's and u's. Since both of these systems would be solvable for the x's in terms of u's, if one equated the solutions for the x's from both systems then, since the two original systems are not identical, one would get a set of equations holding purely among the u's. This implies that some u's are functions of each other which contradicts the functional independence of u's, assumed by the theorem, so the result follows.

Recall the dictum 'no causes in; no causes out'. In keeping with this stance, I look for causal constraints that can be added. The natural one at this stage is to insist, as in ch. 12, that each u_i be a cause of the x_i with which it is mated; that is, that in the system of correct causal laws, there is a genuine causal law for x_i in which u_i appears as cause. In this case the theorem can be of help, though only for very special kinds of system.

Introducing further internal variables (y's) and external variables (w's),[13] consider a set of variables $X \cup Y \cup U \cup W$ for which there is a subset of causally correct equations from a linear deterministic system of the form

$$CL: \quad x_i = \Sigma a_{ij}x_j + \Sigma b_{ij}y_j + u_i$$
$$y_i = \Sigma c_{ij}x_j + \Sigma d_{ij}y_j + w_i$$

where the members of $U \cup W$ are not functions of each other. Suppose there is a set of functionally true candidate causal laws on offer of form CCL over $X \cup U$, where each u_i is mated with the same x_i as in CL. By hypothesis there is one way of writing a set of equations over $X \cup Y \cup U \cup W$ that satisfies the antecedent of the theorem; the theorem shows that there is then only one way. Since the candidate causal laws satisfy the antecedent of the theorem they must be causally correct.[14]

Clearly using this result to infer causal laws puts heavy demands on our antecedent causal knowledge. One important warning to keep in mind is that it is necessary that the underlying correct causal laws have the right characteristics, not just that the candidate laws on offer do. For example, consider again system (1). Suppose that α is indeed a genuine cause of M, β a genuine cause of P, and α and β are not functions of each other. This looks good – but it is no guarantee that the equations of system (1) are causally correct. For these conditions are met if the correct causal laws are instead

$$M \overset{c=}{} \alpha \qquad\qquad\qquad\qquad\qquad (1')$$
$$P \overset{c=}{} a\alpha + \beta$$

In this case the correct causal laws do not satisfy the condition that there be an external variable unique to each equation. Hence the theorem does not apply and we cannot infer that system (1) is causally correct.

The amount of information needed to make use of the result may seem overwhelming. What could enable us to suppose that there was an underlying set of causally correct equations of just the right form, with one factor peculiar

---

[13] The y's are introduced in the equations to cover the possibility of other internal variables appearing in the true causal laws for the x's, while the w's are introduced as external variables for the y's. The value of the inferential results presented (see theorem 2 below) is that it shows that under certain conditions, one can infer from true functional equations containing just x's that the y's above cannot appear in the true causal laws which have form CCL.

[14] So all b_{ij} must be 0.

to each equation, where these factors were just the same as the external variables in the candidate equations? And what could justify the assumption that these special factors were not functionally dependent on each other?

A similar result expressed in a more user-friendly way shows that the antecedent information required is not as daunting as it might first appear. It employs a notion introduced in Cartwright (1989),[15] that of an open back path:[16]

OBP: Given a linear deterministic system, $\langle V, CL \rangle$, where
$V = \{x_1, x_2, \ldots, x_n, y_1, \ldots, y_m, u_1, \ldots, u_{m+n}\}$, x_i has an open back path with respect to $\langle V' \subseteq V \rangle$ if and only if
  (i) there is a law in CL of the form $x_i = \Sigma a_{ij}x_j + \Sigma b_{ij}y_j + u_i$ such that
      (a) any effect of u_i on $x_{j \neq i} \varepsilon V'$ is via x_i
      (b) no $x_j \varepsilon V'$ causes u_i
  (ii) if u_i has causes in V, it has a cause, call it u_i', that satisfies (a) and (b) as well
  (iii) and so on until u_i^m has no cause in V
  (iv) the last u_i^m of clause (iii) – the external variables of CL – are not functions of each other.[17]

Consider a functionally correct candidate causal law for a given effect x_i:

$$\text{CCL}\,(x_e): x_e = \Sigma a_{ej}x_j + u_e$$

Denote the set of variables in CCL(x_e) by V' and suppose that the situation is governed by the 'true' linear deterministic system $\langle V, CL \rangle$ where $V' \subseteq V$. The following theorem follows from the proof in Cartwright:[18]

Theorem 2: if every right-hand-side variable of CCL(x_e) has an open back path with respect to V' and the set of right-hand-side variables contains no factor from any of these open back paths (except u_e) then CCL(x_e) is causally correct.[19]

There is an alternative that looks less complicated but is weaker. Rather than the notion of an open back path it involves the notion of causal connection, which is used frequently in the philosophical literature: two quantities are causally connected if and only if either one causes the other or they have a cause in common.

[15] For a discussion of how this definition of 'open back path' compares with the original in Cartwright (1989) see the appendix to this chapter.
[16] Cartwright (1989), p. 33.
[17] Recalling that for a linear deterministic system all true functional relations must be derivable from correct causal laws, it can be seen that clauses (ii), (iii) and (iv) ensure that the u's in any set containing exactly one factor from the open back path of each variable in V' will not be functions of each other.
[18] 1989.
[19] My original open back path definition did not explicitly require that there be a causal law for for x_i with u_i in it but just that u_i be a cause of x_i. Thanks to Damien Fennell for pointing out to me that, under a natural reading of 'is a cause of', the theorem is open to counterexample without the stronger definition. See the appendix for details.

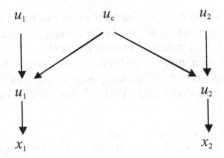

Figure 13.1

Theorem 3: suppose there is a linear deterministic system $\langle V, \text{CL} \rangle$ such that $V' \subseteq V$ and for every right-hand-side variable x_i of $\text{CCL}(x_e)$

(i) there is a law in CL of the form $x_i = \Sigma a_{ij}x_j + \Sigma b_{ij}y_j + u_i$ such that

    (a) any effect of u_i on $x_{j \neq i} \; \varepsilon V'$ is via x_i

    (b) no $x_j \; \varepsilon V'$ causes u_i

(ii) the u_i''s are all causally unconnected.

    Then $\text{CCL}(x_e)$ is causally correct.

Theorem 3 is weaker than theorem 2 because theorem 3 does not apply to candidate systems involving the variables x_1, x_2, u_1, u_2, supposing the true relations are as depicted in fig. 13.1. By contrast theorem 2 will apply so long as u_1' and u_2' have open back paths.[20]

Return to the simple sample systems (1) and (2) for illustration of how theorem 2 can be used for causal inference:

- Suppose we are able to assume that $V' = \{M, P, \alpha, \beta\}$ is a subset of a set of variables V from a linear deterministic system $\langle V, \text{CL} \rangle$ with probability measure over $U = \{\alpha, \beta\}$.
- On offer are a set of functionally correct equations over V' in the form of a linear deterministic system:

$$M \overset{c}{=} \alpha \tag{1}$$
$$P \overset{c}{=} aM + \beta$$

- Suppose we are able to assume that the variables in U are independent in the mean in all combinations: $\langle \alpha\beta \rangle = \langle \alpha \rangle \langle \beta \rangle$

[20] This diagram also illustrates that theorem 2 applies more widely than do results, like many of those for Bayes-nets methods, that are restricted to variable sets that are causally sufficient, that is sets for which 'all' common causes of variables in the set are in the set as well (where what 'all' means has to be spelled out carefully to avoid making a condition so restrictive that it cannot apply to systems where causal processes are continuous). In fig. 13.1, the set $\{u_1, u_2, x_1, x_2\}$ satisfies the open-back-path condition (supposing u_1' and u_2' have open back paths) but it is not causally sufficient.

- Suppose we are able to assume that α has an open back path with respect to $\{\alpha, M\}$ and M and β have open back paths with respect to $\{M, P, \beta\}$.

In this case we can conclude that the equations of system (1) are each causally correct and the system is identifiable.

There are two important things to keep in mind that restrict the usefulness of the method. First the method – as always – is geared to finding special kinds of causal relations, those of a linear deterministic system, and it is inapplicable when the causal relations at work or the ones of interest are not of this kind. Second, we need a particular kind of background knowledge to apply it, knowledge of factors that have open back paths with respect to the quantities of interest.

The need for this kind of information is well known from the placebo problem in medical trials, where it is important that the treatment be administered in a way that does not affect the result other than through the treatment. This means that the method of administering the treatment must have an open back path with respect to the result. Sometimes the design of the experiment provides good reason to be confident that this is the case. In other kinds of situation knowledge of the spatial and temporal relations between the quantities can warrant this assumption; in others the warrant can be provided by knowledge of the structure of the situation. So the constraints, though difficult to meet, are not impossible.

For examples in economics we can look to the recent movement to use 'natural experiments' to infer causes from econometric equations. In many of these cases what we do is to pinpoint a variable that we can feel confident does not influence the putative effect other than via the putative cause. Since we do not know for sure whether the right-hand-side variables of the econometric equation are genuine causes or not, it is necessary to be able to suppose that the selected variable does not cause any of these either except via the putative cause. Hence the selected variable must be part of an open back path for the right-hand-side variables. For instance, in James Hamilton's investigation of the effect that a change in the money stock has on interest rates,[21] he treats non-borrowed reserves as having an open back path with respect to interest rates.[22]

Although the epistemic requirements for the open-back-path method are demanding, what is nice about it is what we do not need to know *ex ante*. In the probabilistic theory of causality, which is the philosophical underpinning

---

[21] For a philosophical discussion of natural experiments in economics and of Hamilton (1997), see Reiss (2003).

[22] Specifically Hamilton uses forecasting error as an open back path for non-borrowed reserves; he states 'the error the Fed makes in forecasting the Treasury balance matters for the federal funds rate only insofar as it affects the supply of reserves available to private banks' (Hamilton (1997), p. 82).

for Granger causality, to determine if x_i causes x_e, we need to look at the partial conditional probability of x_e given x_i, holding fixed a full set of other causes of x_e. This means that we must know what such a set consists in (or figure out some way to finesse our lack of knowledge). This is not so with the open-back-path method. Or consider an alternative way to use Simon's ideas. We suppose that the set of right-hand-side variables in $CCL(x_e)$ contains a full set of causes of x_e but may contain more. We then identify the coefficients to find out which are zero, hence which among the right-hand-side variables are genuine causes. This requires that we know *ex ante* that the right-hand-side variables do include a full set of causes of x_e. Again this is not necessary for the open-back-path method.

Nevertheless situations where there is on offer a functionally correct equation where the right-hand-side variables do have open back paths are clearly not all that common. Even less common are those where we have good reason to believe they have open back paths. It should not be surprising though that the constraints are difficult to meet. The method outlined here is what in ch. 3 I called a narrow-clinching method. Since the results follow deductively from the premises of the method, whenever we are confident that the method has been applied correctly, we can be certain of the conclusions. That is a tall order for a scientific method and we should not be surprised that the methods that satisfy it are heavily restricted in their range of application.

The overall conclusion is that we can get causes from probabilities following methods inspired by the work of Herbert Simon. But we should not expect to be able to do so except for special situations, special both with respect to the causal relations involved and with respect to the background knowledge we have of them.

13.1   Cautionary lessons for econometrics

The slogan, 'no cause in; no causes out' needs to be taken seriously. Causal inference requires antecedent causal assumptions and cannot otherwise be valid. Econometricians in my experience hate making assumptions, so much so that they often give up altogether on making causal inferences about the world. I take it that this is, for instance, the reason why James Heckman says that causality is all in the mind or that Christopher Sims, seeming to despair of settling on the right assumptions about other causes that might be affecting the process under study, has urged reporting the results under all plausible alternatives.

Econometricians are also in my experience especially wary of assumptions they cannot test by econometric methods; and this is not unreasonable. When it comes to one's own work and the conclusions it implies, one wants to be able to police the assumptions, to have confidence that they are well warranted. And that will be hard when one does not have a mastery of the methods used for warranting, when the methods come from a different discipline or different branch of one's own, where one cannot see for oneself that the fine details have been carried out properly – indeed they may seem opaque, even dumb – and one does not know the sociology of the field providing the assumptions, who is a really superior practitioner likely to have done it correctly and who is not. Nevertheless, causal inference does require assumptions, heavy assumptions, and many of them may well not yield to direct econometric testing. So considerably more serious interfield and interdisciplinary understanding seems in order.

In this book we have turned up a number of different kinds of assumptions that are frequently required for causal inference from statistics. These include:

- Assumptions about the relationships between causes and probabilities or between causes and functional relations. (These were highlighted in the discussions of Bayes nets and the axioms for linear deterministic systems in part II. They also matter to Granger causality, where it is assumed that once other reasons for a probabilistic dependence between two factors have been eliminated, any remaining dependence must be due to a causal connection, an assumption also made in Bayes-nets methods.)
- Assumptions about what other factors might be affecting the outcome under study. (These are central in experimental methods and methods based on a Suppes-style probabilistic account of causality, such as Granger causality or Bayes nets.)
- Assumptions about the structure (or form) of the correct causal principles we are trying to learn about. (These have been highlighted in the discussion of the Simon-inspired methods discussed above.)
- Specific causal information about specific factors. (For instance, as we have seen above, the information that one particular factor is a genuine cause of another, or that a set of factors have open back paths or that some set of causal factors, like the u's in the theorems here, are variation free.)
- Assumptions about what kind of causal relationship is at stake and assumptions to the effect that all the causal information (like the varieties of information listed above) that is needed for the inference in question is of the right kind.

The need for the first three kinds of assumptions should be clear from the discussions in part II and up to this point in part III. But I should like to spend a little more time on the last. I urged in ch. 2 that there are many different kinds of causal relations. Here is one place where that matters. Much causal inference has to do with form or structure. We saw how form mattered to the invariance results of ch. 9. It is also clearly central in the discussion of Simon here. The Simon-based methods for causal inference that I have presented above really depend on a simple identifiability result: if there is one way of writing a set of linear relations in triangular form with one u peculiar to each left-hand-side variable and no functional relations among the u's, that way of writing the relations, keeping the triangular form and keeping the u's paired with the same left-hand-side variables, will be unique. What turns this result about the identifiability of a set of functional relations of a given form into a result about causal principles are the causal assumptions noted: (1) this is the form of the underlying causal principles we are trying to infer to; (2) the u's in each candidate law are genuine causes of the left-hand-side variable; (3) these causes are variation free.

Now though we must ask 'what kind of causal relations are at stake?' If different kinds of causal relations can be represented in equations of exactly the same form, we had better be clear that the underlying principles that describe just the kind of causal relations we are interested in have the form we are supposing and moreover that the u's in our candidate laws are causes of that kind. Moreover, we must keep in mind my worries throughout about the connection between warrant and use. Once we have legitimately inferred to causal principles of a given kind, we must not lose track of what kind that is; all our further inferences, inferences for prediction, policy and planning, must be legitimated by the particular kind of causal relations we have inferred. We will see a simple example of this with the lemonade/biscuit machine of ch. 14. There we can infer from looking at the blueprint of the machine that P causes M. (In ch. 14 this is called 'production' causation.) But we must not from this make the further assumption that if we fix P that will affect M but not the reverse, for exactly the opposite is the case for the machine in question.

I said above, 'if different kinds of causal relations can be represented in equations of exactly the same form, then . . .' Can this happen? Certainly. The simple linear deterministic systems of this chapter and the next or of ch. 10 are a case in point. Nothing stops them representing either what in ch. 14 are called 'production' relations or what are there called 'strategy' relations. This matters for both inference and use. I have just adumbrated an illustration about use in the last paragraph. For inference, supposing we are able to use the simple Simon-based methods described in this chapter, we had better know whether we are aiming at principles for production causality or for strategy causality, that the underlying structure for that kind of causality has the right form, that the u's are not only causes but causes of that kind, etc. Nor are these the only kinds of causal relations that can take this same form.

The discussion of strategy versus production relations shows that causal principles of very different kinds can have the same abstract form. Nor do strategy and production relations exhaust the possibilities. For yet another example consider the 'underlying structure' that gives rise to a set of one or another of the kinds of causal relations we have already discussed. In some case it is reasonable to ascribe causal relations at this level too, and they may turn out to have exactly the same abstract form as the strategy or production relations. Recall

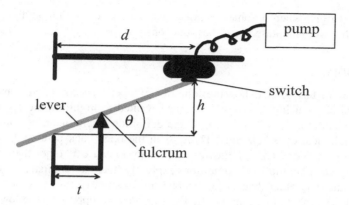

Figure 13.2

the lemonade-and-biscuit machine mentioned above and discussed in detail in ch. 14. Suppose the correct production relations are as represented in equations (2.1) in ch. 14, i.e. pressing on the lever – action α – starts the pump that in turn contributes to turning on the motor. How does α start the pump? Here is one design for bringing it about that pressing the lever starts the pump.

Suppose the lever is attached to a fulcrum inside the machine. Given the depth t of the fulcrum from the wall of the machine, the location of the bottom of the opening in the wall fixes the maximum angle θ through which the lever can rise. When the lever is pushed down, the end of the lever trips a switch closing a circuit that turns on the pump. The switch is affixed to a rigid rod attached to the wall of the machine. Its exact position along the rod is fixed manually when the machine is put together. The location of the bottom of the opening (and hence θ) is also fixed manually at the time of construction. If the switch is located at a depth d from the wall of the machine and height h from the bottom of the opening such that $d \tan \theta = h$, pressing the lever all the way down will trip the switch and turn on the pump. So, let $P \equiv \ln h$, $M \equiv \ln d$, $\alpha \equiv \ln d$, $\beta \equiv \ln \tan \theta$ and represent the fixing of the distance of the switch along the rod by d. Then for the machine to operate properly the rod must be affixed at the height such that the following 'construction'–describing equations are satisfied.

$$M = \alpha$$
$$P = M + \beta$$

But these have exactly the same format features as equations (2.1) in ch. 14, including the fact that α and β are variation free.

The lesson for econometrics is the same as always. Causal inference is hard. It requires a great many antecedent causal assumptions and the causal assumptions have to be about just the kind of causal relation at stake. But it is no good pretending that when it comes to policy we can make do with anything less than valid, well-supported causal conclusions.

APPENDIX

The question has arisen of how the definition of 'open back path' here compares with the original in Cartwright[23] where I wanted to highlight the epistemic requirements for causal inference so the definition was a blend of epistemic and ontological considerations. There u_i in the open back path of x_i on the right-hand-side of $CCL(x_e)$ must be known not to cause x_e other than via x_i. This was taken to imply two conditions explicitly stated here. First, u_i does not in fact cause x_e other than via x_i; second, u_i does not cause any other right-hand-side variable other than via x_i. The second was supposed to follow from the definition because it was supposed that if $CCL(x_e)$ really is a candidate causal law, then no right-hand-side variables are known not to cause x_e. Hence u_i cannot be known not to cause x_e other than via x_i unless it is known that u_i does not cause other right-hand-side variables other than via x_i.

The second place where the epistemic formulation entered was at the last step of the proof. The theorem in Cartwright[24] did not say as here that, given that the open-back-requirement is met, $CCL(x_e)$ is correct if no factor from an open back path appears in it. It said rather that if $CCL(x_e)$ is a candidate for being a correct causal law and the open-back-path requirement is met, it is correct. This is because I took it as known that linear transformations of causally correct equations can produce equations with spurious right-hand-side factors in them, and the appearance of a factor from an open back path is a symptom of a linear transformation. So if a back-path factor appears then it is not the case that for all we know $CCL(x_e)$ is a correct causal law; hence it is not a candidate causal law.

There is a different kind of change here, due to a problem pointed out by Damien Fennell.[25] In Cartwright[26] the definition of an open back path was formulated in terms of x_i having a cause u_i . . . Here it is required that there be a causal law for x_i with u_i in it. This is because if we adopt the transitivity of causation, u_i can indeed be a cause of x_i in cases where it causes, say, y_1 and y_2, both of which cause x_i but the influence of u_i via y_2 cancels its influence via y_2. In that case there will be no causal law for x_i involving u_i and the proof of the theorem requires that there be a law for x_i in which u_i appears.

[23] 1989.   [24] 1989, p. 37.   [25] See Fennell 2005a.   [26] 1989.

14   The merger of cause and strategy: Hoover on Simon on causation

14.1   Three theses

When Kevin Hoover[1] analyses the work of Herbert Simon on causation in deterministic systems, Simon turns out to be describing a far different kind of causal relation than when I study Simon. Mine are the more conventional relations to study in the philosophical literature but Hoover's are more readily of use for policy – and hence might reasonably be at the core of concern in economics. Hoover's are also more widely applicable, though we may more often lack the information necessary to establish them. So, I shall argue for three theses here:

1 Hoover (with macroeconomics in view) studies a different kind of causal relation from 'usual' (for instance, different from most of the others described in ch. 2 in this book). He studies strategy relations rather than production relations.[2]

2 Knowledge of strategy relations is more immediately helpful for setting policy than is knowledge of production relations.

3 If we demand that production relations be able to deliver exactly the same kind of advice for policy as strategy relations, then production relations will only obtain in a subset of the systems in which strategy relations obtain. They may though be easier to learn about.

I should note that in keeping with my own treatment of Simon, I shall assume throughout that causation is asymmetric – quantities do not mutually cause each other. Hoover objects to this assumption in studying macroeconomic quantities and his arguments have convinced me that he is right. Among other things, he points out that macroeconomic variables are often measured over long periods of time and that these long-period variables will mutually affect each other. Nor is it always a solution to try to shorten the time periods so that causal relations follow the usual temporal ordering; this frequently may not even make sense.

This paper is a development of a lecture given in the Economics Department at the University of Birmingham and I am grateful to the audience there for helpful discussion.
[1] Hoover (2001). This is what I say Hoover's definitions, as stated, imply. He does not fully agree.
[2] Though see section 14.5 below where I point out that 'production' is a misnomer.

For instance, he asks, does GDP go to zero for a period if all the factories close at night?[3]

If we do not rule out mutual causation, though, we already start with a glaring difference between Hoover's causal relations and what I am calling 'production' relations. So I will forbid it here since that makes Hoover's causal relations look more like production relations to begin with so that the other differences I want to point to will be clearer. It will also allow for a simplification of Hoover's definitions.

14.2   What is strategy causation and why is it good for policy?

Consider the simple claim, 'x causes y'. A production account of causation focuses on the relation between x and y: x produces y; x makes y happen or is responsible for y; y comes out of x. Hoover's strategy account by contrast focuses on the relation between us and x, y: we affect y by/in affecting x. So for strategy causation we do not consider what happens to y by virtue of what x does but rather what happens to y by virtue of what we do to ensure that x happens. Roughly, x strategy-causes y if and only if what we do that is sufficient for the value of x to be fixed is 'partially sufficient' for y to be fixed.

Formally, Hoover divides the quantities of interest in a given situation into three categories represented by three different kinds of variables:

1 A set of field variables ($\mathcal{F}$) : members of $\mathcal{F}$ remain constant over the period and domain of interest.

2 A set of *parameters* ($\mathcal{P}$) : members of $\mathcal{P}$ represent quantities whose values we can directly control. 'Direct control' is a primitive notion in Hoover's account. It implies at least that the quantities represented by[4] parameters are not caused by any quantities represented in the system. He also takes it to imply that the parameters are variation free: that is, they can take in combination any values allowed for each.

3 A set of 'variables'($\mathcal{V}$) : members of $\mathcal{V}$ represent all the other quantities of interest.

A causal system for Hoover is an ordered triple $\langle \mathcal{F}, \mathcal{P}, \mathcal{V} \rangle$ and a mapping from $\mathcal{F} \times \mathcal{P}$ on to $\mathcal{V}$, that is, a mapping that assigns values to all the variables for each of the allowed values of $\mathcal{F} \times \mathcal{P}$. So relative to an assignment of values to the field variables, the parameters are sufficient to fix the values of the variables. Since the field variables are supposed to be constant over the period of interest, in keeping with Hoover's practice, I shall henceforth suppress mention of them.

For any $v \, \varepsilon \, \mathcal{V}$, define $P_v =$ the minimal set of parameters sufficient to fix the value of v. Now (simplifying Hoover somewhat) we can define:

---

[3] Hoover (2001), pp. 135–6.
[4] I will henceforth drop the distinction between quantities and their representations except where it might lead to confusion.

For $x\,y$ in $\mathcal{U}$, x strategy-causes y if and only if $P_x \subset P_y$.

That is, x causes y if and only if what we do to fix the value of x partially fixes the value of y but not the reverse.

We should notice that it takes a causal notion – that of direct control – to characterize what a parameter is, but from then on the relations necessary to characterize strategy causation are all functional. One might hope that something weaker would do, for instance, that parameters could simply be characterized as a variation-free set. But Damien Fennell shows that this will not work,[5] even if we add that parameters must not be caused by anything in the system. That is because the same set of functional relations can be expressed with different choices of a set of variation-free parameters, giving rise to contradictory causal judgements – x causes y using one parameter set and y causes x using the other. What the notion of direct control does is to tell us which set of variation-free parameters is the right one for characterizing causal relations in the system.

Given that direct control seems a strong notion, we may ask if it is likely to be useful in practice. Hoover thinks the answer is 'yes'. For instance he argues, 'No macroeconomic theory disputes the ability of the Federal Reserve to use its ability to supply or remove reserves from the banking system to set the level of the Federal funds rate' (p. 125). Also,

Every macroeconomic theory that I know predicts that actions that increase the general price level or the Federal funds rate will shift the yield curve upwards in the short run. And at least if the changes are unanticipated, increases in the general price level, will reduce the level of the real interest rate. The empirical evidence for these effects is overwhelming.

I am inclined to agree with Hoover, particularly since I argue that there is a great variety of causal notions that are not reducible to functional ones. I see no reason why the idea of what we can control, and can control without acting via some particular specified set of quantities (the variables of the system under study), may not be among them.

The truth of my second thesis should be obvious now. Following Hoover's example, suppose we want to increase yield. Knowing that increasing the Federal funds rate strategy-causes upward shifts in the yield curve gives us an instrument to accomplish just what we want. For what strategy causation tells us is that anything we do that increases the Federal funds rate will indeed partially ensure that the yield curve goes up. Of course there are still issues of the kind I discuss in ch. 16 to be faced about knock-on and side effects and about costs and benefits. But the strategy-cause information is a valuable asset.

I turn then to my first thesis, which is really the central point I want to make. It should be apparent from the definitions that whether x causes y in Hoover's

---

[5] Fennell (2005b). See also pp. 192–3 above.

account does not depend on the way the equations linking x and y are written. So different ways of writing the equations, which would normally be taken to suggest different causal arrangements between x and y, will all have only one Hoover-cause arrangement. In general though the different ways of writing the equations might all reflect possible production orders – different possible ways the mechanism under study (be it a machine or a bit of the economy) might be operating. This is the idea I will now exploit to illustrate how production and strategy relations differ.

14.3   Two examples where strategy and production come apart

Hoover compares his treatment of causality with J. L. Mackie's INUS account,[6] where x is an INUS condition for y just in case x is an *i*nsufficient but *n*ecessary part of an *u*nnecesary but *s*ufficient condition for y. The comparison is apt. For Hoover, x causes y if everything we do that sets the value of x is part of a condition sufficient for the value of y to be set. But doing what is necessary to set x is not enough for setting y since the parameters of x are a proper subset of the parameters of y. So setting x is insufficient for y though it is a necessary part of what is sufficient since there is no way for the parameters of y to be set without setting those of x.

Mackie, however, does not take the fact that x is an INUS condition for y to be enough to ensure that x causes y, where I think he is talking about production causation. That is because there are well-known counterexamples, two of which we shall look at in this section. Mackie himself adds a condition to the effect that x fixes y, where 'fixes' is a kind of primitive causal notion. Hoover too makes an addition. For him it must be possible that we fix the parameters that set x and y.

The difference here is just the one I described at the start. Mackie, with I believe production causation in mind, attends to a first-level relation between x and y: x fixes y. Hoover, however, tends to a relation between us and x, y: we fix y in part by fixing x. This is why his relations are not production relations. Viewed as such, they are subject to many of the same kinds of counterexamples that beset a simple INUS account of causality that leaves out Mackie's special extra relation between a cause and its effect.

Two well-known kinds of case illustrate that production and strategy causation are different. The first is the problem of joint effects and the second, the equivalence problem, where alternative but equivalent ways of writing functional relations suggest contradictory causal orders.

---

[6] Mackie (1974).

Figure 14(1.1)

Figure 14(1.2)

14.3.1   Joint effects

Consider equation system (1.1) and the corresponding fig. 14(1.1), where the arrows in the diagram represent your favourite kind of production causality. For these purpose I always think about mechanical devices like bicycles or toasters, where production causality depends on the operation of simple machines. (I use Greek letters for parameters, Latin letters for variables and suppress the field variables, which are meant to stay constant.)

$$M = \alpha \tag{1.1}$$
$$P = \alpha + \beta$$

So here α is a joint production-cause of both P and M and β contributes to the production causation of P but not of M.

Despite the fact that we have supposed that there is no causal connection in our machine between P and M, that is not how it looks from the strategy-point of view. Since $Par_M = \{\alpha\} \subset Par_P = \{\alpha, \beta\}$, M strategy causes P. These relations are represented in fig. 14(1.2), where the thick arrows indicate strategy causation, and in the corresponding equations (1.2):

$$M = \alpha \tag{1.2}$$
$$P = M + \beta$$

So strategy causation and production causation are not the same.

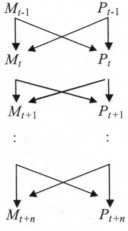

Hoover's fig. 6.1

Figure 14(2.1)

14.3.2   The equivalence problem

In his discussion of temporal order in economics, Hoover gives the example of
the simple dynamic system we see in his fig. 6.1,[7] where the arrows represent
'short-run causality'. In this diagram M and P at t jointly cause M and P at
$t + 1$ and so forth. The diagram corresponds to the following 'error-correction
model' (p. 139):

$$\Delta P_t = \Delta M_{t-1} + \beta(M_{t-1} - P_{t-1}) + \delta + \varepsilon_t$$
$$\Delta M_t = \Delta P_{t-1} + \mu M_{t-1} + \tau + w_t$$

---

[7] Hoover (2001), p. 139.

The equilibrium, or 'long-run' solution for this model is given in equations (2.1) (where I have changed the parameters for simplicity of notation):

$$M = \alpha \qquad\qquad\qquad\qquad (2.1)$$
$$P = M + \beta$$

So Hoover himself acknowledges different types of causality – the 'short run' and the 'long run'. He tells us (pp. 139–140),

Despite the absence of temporal order in the steady state, M causes P in Simon's [NC: 'strategy'] sense. This seems a natural sense of 'cause'. In the short run a change to any of the parameters . . . compels a change to both M and P in a well-defined temporal succession. But the steady-state value of M cannot be affected by the setting of . . . [β], while any change to its own steady-state value due to a change in . . . [α] forces the steady-state value of P to change.

This, however, is not the difference I am pointing to. For in Hoover's case the different kinds of causal relations hold between what are in fact different quantities – the value of M or P at each particular time versus the value of M or P in equilibrium. I want to point out that even when the very same quantities are at stake, there are still different kinds of causal relations to be considered.

I will use Hoover's own equations (2.1) to illustrate. Hoover in this case obviously has in mind (equilibrium) price and (equilibrium) money as the quantities of interest. For these perhaps the idea of production causation does not make sense so that we may be left with strategy causation as the only interesting concept of causation that could apply. To show clearly that where both apply they can be distinct I shall use instead an example where P and M represent quantities that can clearly have both kinds of relations – a case where we have a machine, a real mechanical machine, that we manipulate to produce desired results. The example comes from my visit to the Institute of Advanced Study in Bologna.

There was a lovely machine in my residence in Bologna that dispensed lemonade and biscuits. When it dispensed lemonade it made clack-clack noises – by a pump, I was told; for biscuits, a whirring noise, by a motor. Most often it made both kinds of noises and gave out both lemonade and biscuits. I never knew if the motor tripped the pump or the reverse or neither, though I was told that the whole thing was made from bicycle parts by local students so I knew that whatever connections there were, were all mechanical. There were levers on the machine to push in order to get the lemonade and biscuits but my Italian was not good enough to read the instructions. So whichever I wanted, I just pulled all the levers and I always got both.

This machine can provide a second example of how production causality and strategy causality diverge. My first hypothesis of how the lemonade and biscuits

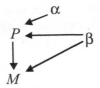

Figure 14(2.2)

were dispensed is represented in fig. 14(2.1), where the quantity of action of the motor is represented by M, the rate of the pump by P, the motion of two levers by α and β. For simplicity I also suppose that all the relations are linear, as represented in the equation set (2.1).

On the other hand I also had the feeling that I heard the pump first and also that the motor whirred less vigorously when the β lever was pulled. So I entertained an alternative hypothesis as well, represented in fig. 14(2.2) and equations (2.2), which keep the same functional relations but with a different causal order.

$$P = \alpha + \beta \qquad (2.2)$$
$$M = P - \beta$$

Later I met the students who built the machine. They told me that my second hypothesis was right: α and β started the pump, which triggered the motor but β had an attachment that damped the motor.

What are the strategy-causal relations for this machine? α and β are parameters sufficient to fix both M and P and

$$Par_M = \{\alpha\} \subset \{\alpha, \beta\} = Par_P$$

so on Hoover's account M strategy-causes P. And that is reasonable. Anything I could do to set P set M as well, but not the reverse, even though the production relations are just opposite to the strategy ones. If we insist that the two must match, it seems the students must have been lying – they could not have built the machine as they said. But that makes no sense. I take it that they built it just as they said but that production and strategy are not the same.

14.4   Modularity – a reconciliation of production and strategy?

There is an immediate solution to the joint effects problem that Hoover might avail himself of. He could take up the claim of Daniel Hausman, James Woodward, Judea Pearl and others discussed in chs. 7 and 8 in this book that we cannot properly call a system causal unless it is modular – that is, unless for each variable v in the variable set $\mathcal{V}$ of the system there is a way for v to be fixed that has no effects in the system other than via v. In Hoover's case where we assume that the laws are not changeable (and assuming linearity through-out), modularity demands that the system must be epistemically convenient, as defined in ch. 10.

This is an extremely strong condition, as I argued in chs. 7 and 8, so Hoover like me may not wish to adopt it as a universal constraint on causality.[8] Worse, though, since facts about causal order are fixed for Hoover entirely by facts about parameters, modularity can only help with the joint effects problem if it is additionally assumed that the special variables that ensure epistemic conve-nience can be counted as parameters – which means that we must be able to control each of them directly – and that is a stronger requirement still. But let us adopt the proposal to see if it works. Suppose then

EC/Par: for every variable v in $\mathcal{V}$ there is a cause u_v of v that causes nothing else in $\mathcal{V}$ except via v, and $u_v \; \varepsilon \; Par_v$.

The idea is that cases like (1.1) do not really exist (or the relations therein are not properly labelled 'causal'); instead the true structure must be that of (1.2). In this case Hoover's analysis yields the results that neither P nor M strategy-causes the other, in exact line with the facts about production causation.

What about the equivalence problem? If the students are not telling the straight story about how they built the lemonade/biscuit machine and (2.1) is the correct production picture after all, the machine satisfies *EC/Par* and the strategy-cause diagram is identical to it. But what if they have indeed built the machine to the blueprint of (2.2) as they say? According to our new assumption, that could never happen. P and M must be epistemically convenient, even if the factors that guarantee this are not salient. The true diagram would then have to be (2.2′), with corresponding equations

$$P = \alpha + \beta$$
$$M = P - \beta + \varphi$$
(2.2′)

So there would have to be somewhere on the machine another lever φ that I never noticed.

Consider now what happens if the same functional relations are assumed to hold but it is M that trips P as the students claim, and not the reverse. In this

---

[8] Perhaps he does though, at least for any situations for which we may be concerned to deliver causal verdicts. For instance he says of an example by Michael Tooley in which modularity fails, 'The situation is pragmatically impossible; there is no situation related to human interest or purpose that is analogous to this one' (Hoover 2001, p. 104). I gather that Hoover thinks this in part because he believes that there are always factors in the causal field that could be 'brought out of' the field and put into the set of parameters. But this proposal needs a great deal of spelling out: what does it mean; will causal order then be relative to what we have decided to put in the field versus the parameter set; will causal order be consistent as variables are successively brought out of the field and put into the parameter set, etc.?

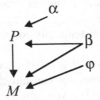

Figure 14(2.2′)

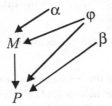

Figure 14(2.1′)

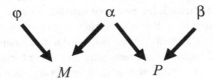

Figure 14(2.3)

case the new equations will be

$$M = \alpha + \varphi \qquad (2.1')$$
$$P = M - \beta + \varphi$$

with corresponding fig. 14(2.1′). Notice that both diagrams blueprint possible machines to build, but with opposite production-cause order for P and M. Clearly the strategy-cause order cannot agree with both. Which does strategy causation favour? Neither. For both equations (2.1′) and (2.2′), $Par_M = \{\alpha, \varphi\}$ and $Par_P = \{\alpha, \beta\}$, which gives fig. 14(2.3) in which neither P nor M causes the other.

Nothing so far assumed stops the building of either the machine represented in (2.1′) or that of (2.2′), with either of the opposing causal orders for P and M; nor does there seem to be anything else that would prohibit either that we have not formally assumed. Yet the strategy-causal verdict about P and M will not coincide with the production-cause arrangement in either case. So it seems we must conclude that strategy causation and production causation are different.

14.5   Is the alternative to strategy really production?

So far I used the label 'production causality' for the relations I am contrasting with Hoover's strategy causation and I have focused attention on machines that embody these kinds of relations. But that focus was far too narrow. Nothing has been assumed in the examples about mechanical causation or any of the characteristics special to it, such as the existence of spatio-temporally continuous causal processes linking cause and effect. What has been assumed is that all the functionally true relations that hold in the given situation can be represented in what in ch. 10 is defined as a linear deterministic system. This is a system of linear equations for which there is a special subset – the 'causal' equations – that generates the rest, where the cause–effect relations represented in the equations satisfy a number of common axioms, such as asymmetry and irreflexivity.

These are certainly weaker assumptions than we would expect for mechanical causation or for any causal-process notion of causality. After all, as noted at the end of ch. 13, strategy relations too can take exactly the same form. It seems thus that my label for these relations has been ill chosen. But it is difficult to find an alternative. On the other hand, an alternative is not really needed for the central point here. Hoover's relations are one special kind of causal relation, one especially geared to use; and verdicts about the kinds of relations Hoover defines must not be taken to warrant claims about any of the other kinds of causal relation we might be concerned with.

14.6   Can strategy relations be rock bottom?

I began this book with the claim that different kinds of causal relations are exhibited in different kinds of system. The contrast discussed here between strategy and production/structural relations is a case in point. Still, one might feel, the strategy relations themselves cannot be basic. They must depend on some other causal relations that look far more like production relations. I think this is an assumption we must be wary of.

Notice Hoover's own derivation of the strategy equations (2.1). Here there is indeed something else – a process that is at least time-ordered – that is supposed to 'undergird' the strategy relations. Recall though that the process he describes in his fig. 6.1 does not involve the same quantities as do the strategy relations, and in particular it is not a process that fills in the intervening steps, starting with the strategy cause (the equilibrium value of M) and ending with the strategy effect (the equilibrium value of P). So if one does have the view that strategy relations must always rest on something else, one will have to be very careful how that view is formulated and just what the 'something else' can be.[9]

A far weaker view would be that there is always some thick(er) description of what is going on, a description that uses causally loaded language, of which the strategy description is a good abstract representation. The difference between the stronger intuition and this is that this weaker view does not suppose that the thick causal facts that support a set of strategy relations must be of some specific kind (or kinds). They need not be representable as any recognizable kind of causal system with fixed characteristics, such as a Bayes-nets system, a linear deterministic system or a spatio-temporally continuous causal process. And if we can reconstruct a particular case formally as a system with particular features, we do not expect that these features must be shared by all cases where strategy relations hold nor do we expect that all the 'underlying' descriptions must have anything interesting in common.

14.7   The range of production relations versus strategy relations

In general it is difficult to compare the range of different kinds of system where different kinds of causal relations apply. As I described in parts I and II, Bayes-nets causation applies to systems that satisfy the causal Markov condition, faithfulness and minimality and for these it may not be possible to fill in temporally intervening variables between any cause and its effect.[10] But just this is required for causal process causation. On the other hand causal process causation can hold where the requirements for Bayes-nets relations fail. Which has the wider range of application?

In our case though a comparison can be made. The causal knowledge obtainable by Simon's methods as reconstructed by me in ch. 13 is knowledge of production relations.[11] For this to have the same kind of immediate causal relevance as Hoover's strategy relations, the systems under study must satisfy

---

[9] For a similar point see ch. 5 in Cartwright (1999). There I consider a version of a causal process requirement: a cause and an effect must always be connected by a spatio-temporally continuous process that 'carries' the causal influence from one to the other. We might insist that this requirement be met in each actually occurring case of a cause–effect relation – that is, on the singular level. But that does not show that at the level of structural/production principles (generic-level causation) there is always between any two principles another principle describing the production of a temporally intermediate effect. What stands between in particular cases may be a myriad of different intermediate processes with different features that fall in no natural category or set of categories.

[10] Cartwright (1999), ch. 5.

[11] Again, this is not really correct; see section 14.5 above where I point out that 'production' is a misnomer.

an additional strong constraint beyond those listed in ch. 13. If this constraint is added, then every system that can be treated using my version of Simon's methods can be treated using Hoover's version, but not the reverse.

To see why consider the immediate policy relevance of Hoover's strategy relations. If the econometrician can assure us that there is good evidence that x strategy-causes y, then we know that if we go about setting the value of x, y will be affected. To get the same information applying my version of Simon's production/structure methods will take a two-step process with two corresponding limits on scope. The econometrician must first determine the production relations. On my reconstruction this will require the system to meet all the assumptions of ch. 13. This means among other things that the relations among the variables in V are identifiable and that each variable in V has an open back path, where the open-back-path variables are in U.[12]

How then is the information about causal relations to be put to use? If we know that x causes y, we can consider changing y by changing x. If we are to do so without disturbing the very principles by which the system operates, we shall have to change one of the u's that causes x, and if we want to ensure that in so doing we are not changing y in any other way, we had best do so by changing a factor in the open back path of x – that is, u_x. But to do that we must be able to control u_x.

Suppose then that we add the constraint that the members of U are all controllable by us. This will severely limit the range of situations to which the account applies. It will also ensure that all of Hoover's conditions are met since then the members of U are sufficient to fix the values of the members of V and they are all controllable by us; that is, the u's are parameters in Hoover's sense. So any system for which we can use Simon's production/structure relations for policy advice in the way suggested[13] will be a system to which Hoover's notion of strategy causation applies.

The converse though is not the case. In a Hoover system, the relations between the parameters and the variables will be identifiable, but the relations among the variables themselves need not be. But identifiability of the relations among the variables is a starting point for the applicability of Simon's production/causality relations as reconstructed by me. So Hoover's strategy concept of causation applies to every system in which my production/structure concept applies if we

---

[12] I should note that Hoover himself puts my claims about the assumptions back to front. I say that one can distinguish mere INUS conditions from genuine (production/structure) causes in a system if the system satisfies the open-back-path assumption. Hoover reports me as claiming that we can distinguish the two only if the system satisfies the assumption (Hoover (2001), p. 103).

[13] There will of course be various other ways to use this information than the straightforward connection suggested: change x to change y. So we must be careful not to read what I say as the strong claim that every time production causation could be useful for policy in any way, we will have a system to which the concept of strategy causation applies.

demand that the latter systems also yield the same type of policy advice we can get from Hoover's strategy relations. But not the reverse.

There are however some drawbacks. Hoover is essentially looking at reduced forms. In the settings he considers, the relation between the variables and the parameters is indeed identifiable – that is, from the probability distribution of $P \cup \mathcal{V}$ we can ascertain the relations between parameters and variables. But how will we learn this probability distribution? Given the nature of what Hoover calls parameters, it is unlikely that we will observe sufficient variation to estimate this. This is a standard problem in dealing directly with the reduced form.

When we consider production relations, however, we are looking at the relations among familiar economic quantities. For these we often have a great deal of both theory and data. We often get the reduced form as just that, as – and only as – a reduced form from a set of equations determined partly by theory and partly by econometric identification. This means that we have more help in learning about production causality than about strategy causality.

So when it comes to the kind of more immediate policy relevance that Hoover introduces, Simon's methods as reconstructed by him are more widely applicable than as reconstructed by me. On the other hand, by ignoring the internal production relations among economic quantities, Hoover loses the potential for theory to help in the discovery of the policy-relevant relations he highlights. The lesson seems to be that both concepts of causation should have their place in economics.

15   The vanity of rigour in economics: theoretical models and galilean experiments

15.1   Introduction

My topic in this chapter is the old and familiar one of the unrealism of assumptions in economic models, especially models aimed at establishing causal connections. For a long time I have maintained that economics is unfairly criticized for the use of unrealistic assumptions.[1] I can summarize my view by comparing an economic model to a certain kind of ideal experiment in physics: criticizing economic models for using unrealistic assumptions is like criticizing Galileo's rolling ball experiments for using a plane honed to be as frictionless as possible. This defence of economic modelling has a bite, however. On the one hand, it makes clear why some kinds of unrealistic assumptions will do; but on the other, it highlights how totally misleading other kinds can be – and these other kinds of assumptions are ones that may be hard to avoid given the nature of contemporary economic theory.

The theme for the volume in which this chapter originally appeared is experiments in economics. My project is not to understand experiments but rather to use experiments to understand theorizing in economics; more specifically, to understand one particular mode of theorizing that is prominent in economics nowadays – theorizing by the construction of models for what Robert Lucas describes as 'analogue economies'.[2] Lucas does not define exactly what an analogue economy is. What I have in mind is theorizing by the construction of models that depict specific kinds of economies and depict them in a certain way. We do not in this kind of theorizing simply lay down laws or principles of a specific form that are presumed to obtain in the economy, as we might in setting out a large-scale macroeconomic model whose parameters we aim

Work on this project has been supported by the Measurement in Physics and Economics Project at the LSE and by the Leverhulme-funded project on the Historical School at the Centre for History and Economics, Cambridge. I am grateful for both the financial and the intellectual help from these two groups, as well as to Sang Wook Yi for helping with the last stages of argumentation and preparation. This paper was first presented at a conference on experiments in 1999 (The Fifth Annual European Conference on the History of Economics, Paris). It sat a long time waiting for publication in the volume arising from the conference.
[1] Cartwright (1989; 1998).   [2] Lucas (1981), p. 272.

to estimate. Rather we justify them from our description of the agents, or sectors, or other significant causal factors in the economy and our description of their significant actions and interactions. Economic principles are employed of course, of necessity, such as the demand for equilibrium of some kind, or the assumption that economic agents act to maximize what they take to be their self-interest. But the detailed form of any principles or equations used will be peculiar to the kind of economy described and the kinds of interactions that occur in it.

Analogue economies generally have only a small number of features, a small number of agents and a small number of options for what can happen, all represented by thin concepts. I call the concepts 'thin' because, although they are often homonymous with everyday economic concepts or occasionally with concepts from earlier economic theories, little of their behaviour from the real world is imported into the model. Seldom, for instance, do we make use of 'low level empirical' relations established by induction. Instead, as we shall see, the behaviour of the features they represent is fixed by the structure of the model and its assumptions in conjunction with the few general principles that are allowed without controversy in this kind of theorizing.

Lucas is a good spokesman in favour of this kind of theorizing, and that is why I cite him. But the method is in no way peculiar to his point of view. Modelling by the construction of analogue economies is a widespread technique in economic theory nowadays; in particular, it is a technique that is shared across both sides of the divide between micro- and macroeconomics. It is the standard way in which game theory is employed; the same is true for rational expectations theory and also for other kinds of theorizing that rely primarily on the assumption that agents act to maximize their utility. As Lucas urges, the important point about analogue economies is that everything is known about them – i.e. their characteristics are entirely explicit[3] – and within them the propositions we are interested in 'can be formulated rigorously and shown to be valid' (p. 67). With respect to real economies, generally there are a great variety of different opinions about what will happen, and the different opinions can all be plausible. But for these constructed economies, our views about what will happen are 'statements of verifiable fact' (p. 271).

The method of verification is deduction: we know what does happen in one of these economies because we know what must happen given our general principles and the characteristics of the economy. We are, however, faced with a trade off: we can have totally verifiable results but only about economies that are not real. As Lucas says, 'Any model that is well enough articulated to give clear answers to the questions we put to it will necessarily be artificial, abstract, patently "unreal"' (p. 271).

---

[3] Lucas (1981), pp. 7–8.

How then do these analogue economies relate to the real economies that we are supposed to be theorizing about? Here is where experiment comes into play, ideal experiments, like Galileo's balls rolling down a smooth inclined plane. For a long time I have maintained that experiments like Galileo's are the clue to understanding one way in which analogue economies can teach us about empirical reality. They show us why the unrealism of the model's assumptions need not be a problem. Indeed, to the contrary, the high degree of idealization involved is essential to the ability of the model to teach about the real world, rather than being a problematic feature we had best eliminate. But I will return then to the feature of these models that is generally thought to be unproblematic – their use of deduction. For my overall suspicion is that the way deductivity is achieved in economic models may undermine the possibility I open up for them to teach genuine truths about empirical reality. So in the end I may be taking back with one hand what I give with the other.

As I mentioned at the start, this chapter is about a very familiar topic: the unrealism of assumptions in deriving causal results from economic models. Section 15.2 will put this problem in a somewhat less familiar perspective by identifying it with the problem of external validity, or parallelism, in experiments. Section 15.3 explains why experiments matter: because many models aim to isolate a single causal process to study on its own, just as Galileo did with his studies of gravitational attraction. Using the language of John Stuart Mill,[4] this kind of model aims to establish tendencies to behave in certain ways, not to describe the overall behaviour that occurs. For this job, it is essential that models make highly unrealistic assumptions, for we need to see what happens in the very unusual case where only the single factor of interest affects the outcome. Section 15.4 raises the question of how we can draw interesting and rich deductive conclusions in economics given that we have so few principles to use as premises; section 15.5 answers that often it seems we fill in by relying on the detailed structure of the model. But then it takes back the solace offered in sections 15.2 and 15.3. For in that case the conclusions are tied to these structural assumptions, assumptions that go well beyond what is necessary for Galilean idealization; the results do not depend just on the process in question but are rather overconstrained. This means that Galilean inference to tendencies that hold outside the experimental set-up is jeopardized. So in the end the problems involved in using highly unrealistic assumptions can loom as large as ever.

15.2   External validity: a problem for models and experiments alike

Lucas speaks of the analogue economies of contemporary economic theorizing as stand-ins for experiment:[5]

---

[4] 1836; 1843.   [5] Lucas (1981), p. 274.

One of the functions of theoretical economics is to provide fully articulated, artificial economic systems that can serve as laboratories in which policies that would be prohibitively expensive to experiment with in actual economics can be tested out at much lower cost.

As we know from Mary Morgan, many of the originators of econometrics viewed their econometric models in a similar way, for they thought of situations in which the parameters of their structural models could be identified as situations in which by good luck nature is running an experiment for us.[6]

Francesco Guala too talks about the similarities between laboratory experiments in economics and the kinds of theoretical models I am discussing here.[7] Guala has been studying how experiments work; I have been trying to understand how theoretical models work. We have both been struck by the structural similarities between the two. I am particularly interested in the fact that both laboratory experiments and theoretical models in economics are criticized for the artificiality of the conditions they set up. As Lucas says, the assumptions of our theoretical models in economics are typically 'artificial', 'abstract' and 'patently unreal'.

Thinking about this very same complaint with respect to the laboratory experiments we perform nowadays in economics provides us with a useful vocabulary to describe the problems arising from the unrealism of assumptions in theoretical models – and to see our way around them. When we design an experiment or a quasi-experiment in the social sciences, we aim simultaneously for both internal validity and for external validity. An experimental claim is internally valid when we can be sure that it has genuinely been established to hold in the experimental situation. External validity – or 'parallelism' as economists call it – is more ambitious. For that the experiment must be designed to ensure that the result should hold in some kinds of targeted situations or populations outside the experimental set-up.

It is a well-known methodological truism that in almost all cases there will be a trade off between internal validity and external validity. The conditions needed in order to increase the chances of internal validity are generally at odds with those that provide grounds for external validity. The usual complaint here is about the artificiality of the circumstances required to secure internal validity: if we want to take the lessons, literally interpreted (you should note the 'literally interpreted'– I shall return to it below), from inside the laboratory to outside, it seems that the experimental situation should be as similar as possible in relevant respects to the target situation. But for the former we need to set up very special circumstances so that we can be sure that nothing confounds the putative result, and these are generally nothing like the kinds of circumstances to which we want to apply our results.

[6] Morgan (1990).   [7] Guala (2005).

This is exactly what we see in the case of economic models. Analogue economies are designed to ensure internal validity. In an analogue economy we know the result obtains because we can establish by deduction that it has to obtain. But to have this assurance we must provide an analogue economy with a simple and clear enough structure that ensures that deduction will be possible. In particular we need to make very special assumptions matched to the general principles we use: we must attribute to this economy characteristics that can be represented mathematically in just the right kind of form, a form that can be fed into the principles in order to get deductive consequences out. And this very special kind of dovetailing that can provide just what is needed for deduction is not likely to be provided by conditions that occur in the economy at large, as Lucas and all other theorists using these methods admit. In this kind of theorizing it looks as if we buy internal validity at the cost of external validity.

Nor is the problem confined to the 'thought experiments' we conduct with our constructed models. It also appears in the real experiments we conduct nowadays in economics; and it reveals a significant difference in concerns between economics and many other branches of social science. Experimental economists report astounding confirmation of a number of economic hypotheses they have been testing recently.[8] These experimental economists are also very proud of their experimental designs, which they take to have minimized the chances of drawing mistaken conclusions. Yet, apparently, it is difficult for them to get their results published in social science journals outside their own field, because, referees claim, they have virtually no guarantees of external validity.[9] So the results, it is felt, lack general interest or significance.

15.3   Tendencies and Galilean idealizations

Now I should like to argue that a great many of the unrealistic assumptions we find in models and experiments alike are not a problem. To the contrary, they are required to do the job; without them the experiment would be no experiment at all. For we do not need to assume that the aim of the kind of theorizing under discussion is to establish results in our analogue economies that will hold outside them when literally interpreted. Frequently what we are doing in this kind of economic theory is not trying to establish facts about what happens in the real economy but rather, following John Stuart Mill, facts about stable tendencies. Consider a stock example of mine – a model designed by my colleague Chris Pissarides to study the effects of skill loss on unemployment.[10] What we want to learn from the analogue economy described by Pissarides is

---

[8] Cf. Plott (1991) and Smith (1991).
[9] Conversation with Charles Plott, California Institute of Technology, May 1997.
[10] Pissarides (1992).

not whether there will be persistence in unemployment in the real economy but instead what skill loss will contribute to persistence – what skill loss tends to produce, not what eventually occurs whenever skill loss occurs.

So what I maintain is that the analogue economies described in contemporary economic models look like experiments, where the experimental aspect matters. The models almost always concentrate on a single mechanism or causal process. For example, Pissarides' model studies the effect (if any) of skill loss during unemployment on the persistence of unemployment shocks via the disincentives arising from loss of skills in the labour pool for employers to create jobs in areas where skill affects productivity. The idea is to isolate this process; to study it in a setting where nothing else is going on that might affect the outcome as well. The model is constructed to assure us that whatever result we see is indeed due to the process under study.

Consider the skill-loss model. Loss or not of skill during unemployment is the only exogenous variable. Firms act to maximize profits and only to maximize profits. We can trace through the model to see that the only variation in profits will be due to the number of jobs that firms decide to create in the face of a labour pool containing unemployed workers and to the productivity of the workers hired. For this model we can derive rigorously that unemployment in one period is dependent on unemployment in the previous period if and only if skills are lost during unemployment. It looks as if this model allows us to see exactly what effects loss of skill has on unemployment persistence via the disincentive it creates for job creation.

What can we conclude? Can we conclude that we have learned a fact about skill loss *per se*, a fact we can expect to be generally true, true not just in this analogue economy but in other economies as well? Certainly not if we try to read the conclusion as one about the association between loss of skill and unemployment persistence with some kind of quantifier in front: always, or for the most part, or even sometimes, if there is skill loss in sectors where skills matter to productivity, there will be unemployment persistence. Clearly a good deal else could be going on to offset the effects of skill loss, even to swamp them entirely; indeed we might never see persistence in any case of skill loss, even though the model shows correctly that 'skill loss leads to unemployment persistence'.

This is why we turn to the notion of stable tendencies:[11] in any situation skill loss tends to produce persistence in unemployment shocks. What does this

---

[11] Cartwright (1989; 1998). I have myself defended the importance of tendencies throughout the social and natural sciences, wherever the analytic method is in play (see Cartwright (1989)) and have specifically maintained, possibly incorrectly given the arguments here, that we can learn about them via our formal models (see Cartwright (1998)). Daniel Hausman (Hausman (1992)) in his arguments that economics is a separate but not an exact science also sees tendencies as standard in economic theory.

mean in terms of what actually happens? There does not seem to be any general rule in economic theory that answers, as vector addition does on Mill's account of the tendencies of different forces in classical mechanics. Nevertheless, if economic theory is to aspire to be an exact science, there had better be at least a case-by-case answer. And presumably this answer can in general be generated by the specific model that testifies to the tendency, in conjunction with any general economic theory we are in a position to assume.[12] For the skill-loss tendency, I take it that we assume roughly that in any situation where skills matter to productivity and the decision to create new jobs by a firm is in part determined by its expected profit, unemployment at one period will depend on previous levels of unemployment if workers are thought to lose skills during unemployment and not otherwise, even if this dependency on past levels plays only a small part in determining present levels.

Probably no one thinks we have established even that, though, because economists, like other social scientists, are alert to the possibility of inter-action, as Mill himself warned. In some situations some factors may distort the skill-loss mechanism so much that loss of skill behaves differently in those situations from the way it behaves in our analogue economy. Of course if we are going to avoid manœuvres that are entirely *ad hoc* we shall have to ensure that 'interaction' is given real verifiable content whenever it is invoked. In principle this should be possible since the theoretical model is supposed to lay bare how the process operates in the first place – 'distortions' are judged relative to it.

We can see the general points more clearly by thinking again about the kind of laboratory experiment that aims to establish a tendency claim. Perhaps rather than thinking of economics experiments, which tend to be controversial, we should take an illustration from physics, let us say Galileo's famous experiments to establish the effect of the attraction of the earth on a falling body, one of which is illustrated in fig. 15.1 (Leaning tower).

Galileo's experiments aimed to establish what I have been calling a tendency claim. They were not designed to tell us how any particular falling body will move in the vicinity of the earth; nor to establish a regularity about how bodies of a certain kind will move. Rather, the experiments were designed to find out what contribution the motion due to the pull of the earth will make, with the assumption that that contribution is stable across all the different kinds of situations falling bodies will get into. How did Galileo find out what the stable contribution from the pull of the earth is? He eliminated (as far as possible) all other causes of motion on the bodies in his experiment so that he could see how they move when only the earth affects them. That is the contibution that the earth's pull makes to their motion.

---

[12] Cartwright (1989; 1998; 1999). I have elsewhere (Cartwright (1999)) described a variety of rules for combining tendencies besides vector addition, as well as explaining what we can do with tendency knowledge even when there are no general rules available for combining tendencies.

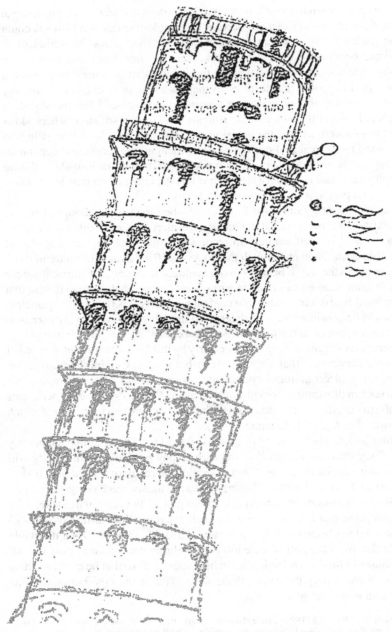

Figure 15.1 Leaning tower (designed by Emily Cartwright)

Let us call this kind of idealization, which eliminates all other possible causes to learn the effect of one operating on its own, Galilean idealization. My point is that the equivalent of Galilean idealization in a model is a good thing. It is just what allows us to carry the results we find in the experiment to situations outside – in the tendency sense. We need the idealizing assumptions to be able to do this. Otherwise we have no ground for thinking the behaviour we see in the experiment is characteristic of the earth's pull at all. Indeed, we know it will not be.

We can contrast these Galilean experiments with experiments that have a quite different aim and correlatively a quite different structure. Consider what happens when we build a prototype of a new device and experiment on it to ensure that it will work correctly when put to use. In this case we do not aim to learn an abstract tendency claim. Instead we want to find out what actual behaviours occur. So the experimental conditions should be very realistic to the conditions in the target situations and vary appropriately across them. And without more said, we have no reason to expect the results in the experiment to obtain in any situations except those that resemble the conditions in the experiment.

Here we see another trade off. If an experiment is very, very unrealistic in just the right way, its results can be applicable almost everywhere. But they will not be able to tell you what happens anywhere else since they only establish the contribution or tendency of the factor in question. Experiments that are very realistic can tell you what happens. But they are highly limited in their scope, for they can only tell you what happens in situations that look like the experimental set-up. And experiments in between are usually pretty uninformative on both matters. Of course we may be very lucky. It may be, for instance, that the cause or small set of causes that we isolate in our experiment (or in our model) is also the dominant cause in the real situations we want to know about. In that case our Galilean experiments (and the corresponding models) will not only give us tendencies but will be approximately descriptively accurate as well.

Back to models again. If the deductions have been carried out correctly and the general principles employed are true in the target situations, the results of the model will obtain in any real situations that fit the description that the model provides. And in general we have no reason to think they will obtain anywhere else. But if what the model describes satisfies the requirements to be a Galilean-style experiment, it can do more. It can tell us what happens in an experimental situation and thereby tell us about the tendency of the features in question. So Galilean idealization in a model is a good thing.

15.4   How deductivity can be secured and at what cost

The problems I worry about arise when not all of the unrealistic assumptions required for the derivations in a model are ones that characterize an ideal experiment. What I fear is that in general a good number of the false assumptions made with our theoretical models may not have the form of Galilean idealizations. Before I go into details about these kinds of extra-Galilean assumptions, I shall first lay the groundwork by explaining why we might expect to find them as features of our analogue economies. The need for these stronger constraints – the ones that go beyond Galilean idealization – comes, I believe, as a result of the nature of economic theory itself. To see how, let us look again at what kinds of theory are available in economics to aid in the construction of models and at what kinds of concepts they deploy.

The bulk of the concepts used in these models are concepts naming socio-economic quantities that are familiar to the layman, not only as the targeted results to be explained but also as the proposed explanatory factors, concepts like persistence in unemployment and loss of skill during unemployment, or current price, tax, demand, consumption, labour, wages, capital, profit and money supply, or assessment of skills, private information and in-firm training, or, to take an example from game-theoretic political economy, power to redistribute, incentives for credible information transmission and political failure in the transmission of information.

This is my first observation: most of the concepts employed in these models are highly concrete empirical concepts. My second observation is that the task is to establish useful relations among these via deduction. The problem comes with my third observation: the theory that is presumed is very meagre. There are not many principles available to use in the deductions. We have only a handful of very general principles that we employ without controversy in economics, such as the principles of utility theory. Nor are there usually many concrete empirical principles imported into the models either. I take it that this is part of the strategy for the models. Almost any principle with real empirical content in economics is highly contentious and we try to construct models that use as few controversial assumptions as possible. But this makes difficulties for the scope of the theory. If the results are supposed just to 'fall out' by deduction from the principles, where there are not many principles, we will not get many results either. How, then, can we deduce results in our models when we have few general principles to call on?

To answer, consider what typical models for analogue economics look like. These models tend to be simple in one respect: they usually have only a few agents with few options and only a narrow range of both causes and effects is admitted. Yet there is another way in which they are complex, at least by comparison with physics models doing the same kind of thing: they have a lot of structure. The list of assumptions specifying exactly what the analogue economy is like is very long. Consider one of Lucas's own models, from his 1973 'Expectations and the Neutrality of Money'.[13] I choose this example because it

---

[13] Lucas (1981), pp. 66–89.

is a paper whose 'technically demanding form' is explicitly defended by Lucas (p. 9). Section 2 is titled 'The structure of the economy', i.e. the structure of the analogue economy that Lucas uses to study money illusion. What follows is section 2 (pp. 67–9) in its entirety:

In order to exhibit the phenomena described in the introduction, we shall utilize an abstract model economy, due in many of its essentials to Samuelson. Each period, N identical individuals are born, each of whom lives for two periods (the current one and the next). In each period, then, there is a constant population of $2N$: N of age 0 and N of age 1. During the first period of life, each person supplies, at his discretion, n units of labor which yield the same n units of output. Denote the output consumed by a member of the younger generation (its producer) by c^0, and that consumed by the old by c^1. Output cannot be stored but can be freely disposed of, so that the aggregate production–consumption possibilities for any period are completely described (in per capita terms) by:

$$c^0 + c^1 \leq n, \quad c^0, c^1, n \geq 0 \tag{1}$$

Since n may vary, it is physically possible for this economy to experience fluctuations in real output.

In addition to labor-output, there is one other good: fiat money, issued by a government which has no other function. This money enters the economy by means of a beginning-of-period transfer to the members of the older generation, in a quantity proportional to the pretransfer holdings of each. No inheritance is possible, so that unspent cash balances revert, at the death of the holder, to the monetary authority.

Within this framework, the only exchange which can occur will involve a surrender of output by the young, in exchange for money held over from the preceding period, and altered by transfer, by the old. We shall assume that such exchange occurs in two physically separate markets. To keep matters as simple as possible, we assume that the older generation is allocated across these two markets so as to equate total monetary demand between them. The young are allocated stochastically, fraction $\theta/2$ going to one and $1 - (\theta/2)$ to the other. Once the assignment of persons to markets is made, no switching or communication between markets is possible. Within each market, trading by auction occurs, with all trades transacted at a single, market clearing price.

The pretransfer money supply, per member of the older generation, is known to all agents. Denote this quantity by m. Posttransfer balances, denoted by m', are not generally known (until next period) except to the extent that they are "revealed" to traders by the current period price level. Similarly, the allocation variable θ is unknown, except indirectly via price. The development through time of the nominal money supply is governed by

$$m' = mx, \tag{2}$$

where x is a random variable. Let x' denote next period's value of this transfer variable, and let θ' be next period's allocation variable. It is assumed that x and x' are independent, with the common, continuous density function f on $(0, \infty)$. Similarly, θ and θ' are independent, with the common, continuous symmetric density g on $(0, 2)$.

To summarize, the state of the economy in any period is entirely described by three variables m, x, and θ. The motion of the economy from state to state is independent of decisions made by individuals in the economy, and is given by (2) and the densities f and g of x and θ.

But this is not an end to the facts set to obtain in Lucas's 'abstract model economy'. Section 3 continues, 'We shall assume that the members of the older generation prefer more consumption to less' and so on for another page; and more details are still to be added to the economy in section 4. There is nothing special here about Lucas though. Just write out carefully in a list the assumptions for almost any of your favourite models and you will see what I mean. For example the skill-loss model of Pissarides contains some sixteen assumptions and that for just the first of six increasingly complex economies that he describes.[14]

I believe there is good reason why economic models must give a lot of structure to the economies they describe: if you have just a few principles, you will need a lot of extra assumptions from somewhere else in order to derive new results that are not already transparent in the principles. In the models under discussion the richness of structure can fill in for the want of general principles presupposed. The general principles can be thought of in two categories, familiar to philosophers of science:[15] internal principles and bridge principles. Internal principles make claims about the relations of abstract or theoretical concepts to each other, like the axioms of utility theory. But the results we want to know about generally involve not abstract or theoretical concepts, but empirical ones. The bridge principles of a theory provide links between the two sets of concepts. (The usual example is the identification in an ideal gas of the theoretical concept mean kinetic energy of the molecules with the empirical concept temperature.)

The theory presupposed in our economics models tends to employ few principles of either category and often no bridge principles at all. This means that the additional assumptions put in via the description of the model must do two jobs. On the one hand they must provide sufficient constraints to serve as premises to increase the range of deductive consequences. On the other hand they must establish an interpretation of the terms that appear in the theoretical principles. They must tell us, for instance, what utility amounts to in terms of an employer's opening a job and of work versus leisure for the employee, or of entrepreneurs investing in a project and of managers defaulting on their contracts, or of fair treatment for one's fellow citizens and of the cost of demonstrating or contributing to the American Civil Liberties Union.

Sometimes the job left open by the want of bridge principles is done by an explicit assumption: we will assume that the only source of utility is . . .

---

[14] Economists, I think, get used to models with lots of assumptions. But I am often talking to mixed groups, people who study economics and people who study physics; those whose background is in physics are often astounded at the richness of description provided in models in economics.

[15] Hempel (1966).

Sometimes the abstract principles themselves are explicitly given a concrete form: we will assume that firms act to maximize profits and labourers to maximize wages . . . Often the interpretation is implicit: perhaps there is nothing else in the model for agents to care about except power or profit or leisure and wages, and the very choice of these words indicates that the agents' utility should depend on them in certain characteristic ways.

My claim then is that it is no surprise that individual analogue economies come with such long lists of assumptions. The model-specific assumptions can provide a way to secure deductively validated results where universal principles are scarce. But these create their own problems. For the validity of the conclusions appears now to depend on a large number of very special interconnected assumptions. If so, the validation of the results will depend then on the detailed arrangement of the structure of the model and is not, prima facie at least, available otherwise. We opt for deductive verification of our claims in order to achieve clarity, rigour and certainty. But to get it we have tied the results to very special circumstances; the problem is how to validate them outside.

Consider for example the Lucas model from 'The Neutrality of Money'. We begin with the fairly vacuous claim:[16]

[T]he decision problem facing an age-0 person is:

$$\max_{c,\, n,\, \lambda \,\geq 0} \left\{ U(c, n) + \int V\left(\frac{x'\lambda}{p'}\right) \, dF(x', p'; m, p) \right\} \tag{9}$$

subject to:

$$p(n - c) - \lambda \geq 0 \tag{10}$$

where c is current consumption; n, current labour supply; λ, a known quantity of nominal balance acquired; p and p', price levels in the current and successor period; and F, an unspecified distribution function. Despite the fact that there is not much that is controversial yet, we can see that even at this stage the exact form of the equation depends on the details of the economy. This is even more obvious by the time we get to the condition for equilibrium in each separate market, equation (16), which is derived from equation (9) plus the more detailed assumptions about the analogue economy studied in the model (p. 72):

$$h\left(\frac{mx}{\theta p}\right)\frac{1}{p} = \int V'\left(\frac{mxx'}{\theta p'}\right)\frac{x'}{p'} \, dF(x', p'; m, p) \tag{16}$$

Sections 6 and 7 of the Lucas paper are entitled, respectively, 'Positive implications of the theory' and 'Policy considerations'. Yet the results he establishes

---

[16] Lucas (1981), p. 70.

are about this economy: they follow from equation (16), which is an equation specific in form to the economy that satisfies the lengthy description laid out in Lucas's sections 2, 3 and 4. How can they teach us more general lessons, lessons that will apply to other, different, economies?

The view that I have long defended is that such model results teach us about general tendencies (in my own vocabulary, 'capacities'), tendencies that are nakedly displayed in the analogue economies described in our economic models but that stand ready to operate in most economies. On this view the analogue economy that Lucas describes is like an experiment. We know that an experiment of the right kind, a Galilean experiment that isolates the tendency in question,[17] can teach lessons that carry outside the experimental situation. If we are lucky, however, we will not need to carry out the experiment. We can find out what would happen were we to conduct it because we can find out by deduction what must happen. But for that to work, the analogue economy must be of just the right kind: were we to construct it in reality, it would meet the conditions of a Galilean experiment. This whole strategy is threatened, however, if non-Galilean idealizations play a role in our deductions – which looks to be the case with Lucas's equation (16).

From the perspective of establishing tendencies, it becomes crucial then to look carefully into the deductions used in our economic models to see if all of the unrealistic assumptions required for the derivations are ones that characterize an ideal experiment. Let us look at another simple physics example for an analogous case.

In classical Newtonian mechanics massive bodies have an inertial tendency: a body will remain in motion unless acted on by a force. When it is acted on by a force the actual motion that occurs will be a combination of the inertial motion and that due to the force. So, what is the natural behaviour of a body when inertia acts on its own? Say we do some experiments to find out. We know that forces cause motions. So eliminate all forces and watch the bodies move. What will we see?

Imagine that our experimental mass has been confined for reasons of convenience to move on a particular surface, but that we have been very careful to plane the surface to eliminate almost all friction. Then what we will see will depend on the geometry of that surface. For example, if all our experiments are done on a sphere, we always get motion in great circles, as in fig. 15.2 (geodesic on the simple sphere geometry). But that is not the 'natural' motion in other geometries. Look for instance at fig. 15.3 (geodesic on the sphere geometry with space–time singularities). There, motion on great circles is available, but it is not the motion that inertia will contribute. The results in our experiment are overconstrained. We thought that by eliminating all the factors we think of as

---

[17] If such a tendency exists.

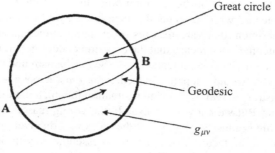

Great circle = geodesic

Figure 15.2   Geodesic on the simple sphere geometry (designed by Sang Wook Yi)

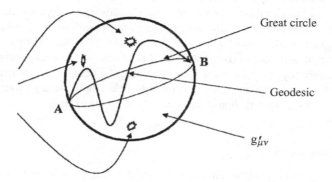

Figure 15.3   Geodesic on the sphere geometry with space–time singularities (designed by Sang Wook Yi)

causes of motion – all the forces – we would see the results of inertia by itself. Instead what we see is a result of inertia-plus-geometry.

This can always happen in an experiment: we never know whether some features we have not thought about are influencing the result. But in a good many of our analogue economies we are not even this well off. In a real experiment we are after all in a position to assume with good justification that the fact that there are, for instance, only two markets or only two generations does not matter because the number of markets or of generations is not relevant to the conclusion: it has no causal bearing on the outcome, and what happens in the real experiment is just what is caused to happen. Analogue economies are different. What happens in them is exactly what is implied deductively. The problem is that we often know by looking at them that the specific derivations made in our models depend on details of the situation other than just the mechanism itself operating in accord with our general principles. So we know that in the corresponding experiment there are features other than the mechanism itself determining the outcome. That means that the experiment does not entitle us to draw a conclusion about the general tendency of the mechanism under study.

We now know what would happen – indeed, what must happen – in some very particular constrained real experimental situation in which the features of interest really occur. But we know it for exactly the wrong reason. We know that the results obtain because we know that they follow deductively given the formal relations of all the factors that figure in an essential way in the proof. But the whole point about an experiment designed to establish the tendency of a factor is that the background factors should not matter to what happens. We are supposed to be isolating the effects of the feature or process under investigation acting on its own, not effects that depend in a crucial way on the background.

So, were such a set-up to occur, it would turn out not to be a good experiment after all. It may have seemed to be a good design because our independent causal knowledge told us that in general none of the background factors should have any bearing on the effect. But by bad luck that would not be true of the particular arrangement of them we chose. The formal relations of the background and targeted feature together are enough to guarantee the result – and that is one of the things our design is meant to preclude. We would have to judge the result (even if by chance it should turn out to be correct) to be an artifact of the experiment.

15.5   Conclusion

Let us look at Lucas's own conclusion in his paper on the neutrality of money:[18]

This paper has been an attempt to resolve the paradox posed by Gurley, in his mild but accurate parody of Friedmanian monetary theory: 'Money is a veil, but when the veil flutters, real output sputters.' The resolution has been effected by postulating economic agents free of money illusion, so that the Ricardian hypothetical experiment of a fully announced, proportional monetary expansion will have no real consequences (that is, so that money *is* a veil). These rational agents are then placed in a setting in which the information conveyed to traders by market prices is inadequate to permit them to distinguish real from monetary disturbances. In this setting, monetary fluctuations lead to real output movements in the same direction.

In order for this resolution to carry any conviction, it has been necessary to adopt a framework simple enough to permit a precise specification of the information available to each trader at each point in time, and to facilitate verification of the rationality of each trader's behavior. To obtain this simplicity, most of the interesting features of the observed business cycle have been abstracted from, with one notable exception: the Phillips curve emerges not as an unexplained empirical fact, but as a central feature of the solution to a general equilibrium system.

[18] Lucas (1981), p. 84.

I have argued that in a model like this the features 'abstracted from' fall into two categories: those that eliminate confounding factors and those that do not eliminate confounding factors but rather provide a simple enough structure to make a deductive study possible. The former I claim are just what we want when we aim to see for rational agents what effects inadequate information about money disturbances has on the short-term Phillips curve, that is, when we want to establish the tendency it has independent of the effects anything else might have on a Phillips curve as well. But the assumptions of the latter kind remain problematic. They not only leave us with the question still unanswered, 'Can we think that what we see happen, literally happen, in this economy, is what the combination of rationality and limited information will *contribute* in other economies?' Worse, they give us reason to think we cannot. For inspection of the derivation suggests that the outcome that occurs in the analogue economy does depend on the particular structure the economy has.[19]

Does it? This is a question that is generally not sufficiently addressed. Frequently of course we do discuss how robust the results from a specific model are. But, not surprisingly, these discussions usually refer to assumptions in the first category, for these are the ones that are of concern to economic theory. Notice for instance that Lucas notes in the passage just cited that 'most of the *interesting* features of the observed business cycle have been abstracted from' (my italics). In the end we want to know what happens when other causes are at work, either because they may interfere with the one under study, or because we are starting down the road toward a model that will be more descriptively accurate when the results are read literally, i.e. more descriptively accurate about the real economies we want to study. But my central point is that we need robustness results about the second category of assumptions as well if our results are to be of use in the tendency sense.

I realize of course that economists do not use models just to find out about tendencies. The models are merely one strand in a net of methods used together for establishing, testing, expanding and revising economic hypotheses. Moreover, it is often the general lesson rather than the precise form of the conclusion that is taken seriously (even when the conclusion is understood in a tendency sense). Nevertheless, rigorously deriving a causal conclusion in a model is supposed to provide prima facie evidence in favour of that conclusion. My concern is about just this relation of evidence to hypothesis. To the extent that the derivation in a model makes essential use of non-Galilean 'idealizing' assumptions, then I do not see how the fact that the result can be derived in such a model can provide any evidence at all for the hypothesis.

---

[19] Notice that we still have this problem even if we are lucky enough to have selected a few causes to study that for most real situations will be the dominant causes. For we still need to see why the very behaviour that occurs in the analogue economy when these causes are present is behaviour that reveals the tendency of this arrangement of causes and hence will approximate the behaviour that occurs in the real economies.

If we aim to establish conclusions interpreted in a tendency sense, there is a good reason why the derivation of a conclusion in a model that makes Galilean idealizations, and no others, should count as evidence in favour of that conclusion: to the extent that the general principles employed in the derivation are true to the world, behaviour derived in the model will duplicate behaviour that would obtain were a Galilean experiment to be performed. But when non-Galilean idealizations are made as well, this reason no longer has force. So we need another reason to show why this procedure has evidential force. And I do not know one that can be stated clearly and defended convincingly. Hence I think we should be concerned to ensure that non-Galilean idealizing assumptions do not play an essential role in our derivations.

What, then, does this tell us about the demand for rigorous derivations? I have here been discussing one of the central and highly prized ways that economic theory is done today: by the construction of models for simple analogue economies in which results about issues of interest can be derived rigorously, employing as general principles only ones whose use is relatively uncontroversial within the discipline. The achievement of rigour is costly however. It takes considerable time. It requires special talents and special training and this closes the discipline to different kinds of thinkers who may provide different kinds of detailed understanding of how economies can and do work. And rigour is bought at the cost of employing general concepts lacking the kind of detailed content that allows them to be directly put to use in concrete situations. What are its compensating gains? Unless we find different answers from the one I offered here,[20] the gains will not include lessons about real economic phenomena, it seems, despite our frequent feeling of increased understanding of them. For we are not generally assured of any way to take results out of our models and into the world.

There has been some tendency to blame our failures on the attempt to make economics rigorous. I am inclined to go the other way. If it is rigour that we want, the problem with economic theorizing of this sort is that rigour gives out too soon. For the models themselves, though abstract and mathematized, are not formal theories. To see why I say this, consider again the structure of my argument in this chapter. I have raised questions about the external validity of the results established in these kinds of models. My worries focus not on the unrealism of the assumptions but on the model-dependence of the results. The kind of model-dependence involved seems to undercut not only the claim that the results can be read literally, but also the hope that they can be read as facts about tendencies.

---

[20] There are of course a variety of other accounts of the use of models that do not demand either predictive accuracy or the correct isolation of a tendency. See for instance the studies found in Morgan and Morrison (1999).

But I have to say 'seems' here because the models themselves are not presented in a way that allows this question to be taken up easily or answered rigorously. What exactly are the assumptions that are really necessary for the derivations to go through; and what is the range of circumstances across which these assumptions can be relaxed and qualitatively similar results still follow? We cannot generally answer that question given the way the models are presented. To answer it we need to formalize our models. Supposing then that my worries about the model-dependence of the results are valid. What should we conclude about the need for rigour in economic theory? It looks as if the natural conclusion is this: should economics stick to mathematizing rather than formalizing, it will not be easy to know whether the models it constructs can teach us general facts about concrete features of the economy or not; the trouble with this kind of theorizing is not that it is too rigorous, but rather that it is not rigorous enough.

16   Counterfactuals in economics: a commentary

16.1   Introduction

Counterfactuals are a hot topic in economics today, at least among economists concerned with methodology. I shall argue that on the whole this is commonly a mistake. Frequently the counterfactuals on offer are proposed as causal surrogates. But at best they provide a 'sometimes' way for finding out about causal relations, not a stand-in for them. I say a 'sometimes way' because they do so only in very special – and rare – kinds of system. Otherwise they are irrelevant to establishing facts about causation. On the other hand, viewed just as straight counterfactuals, they are a wash-out as well. For they are rarely an answer to any genuine 'What if . . .?' questions, questions of the kind we pose in planning and evaluation. For these two reasons I call the counterfactuals of recent interest in economics, impostor counterfactuals.

I will focus on Nobel-prize-winning Chicago economist James Heckman, since his views are becoming increasingly influential. Heckman is well known for his work on the evaluation of programmes for helping workers more effectively to enter and function in the labour market. I shall also discuss economist Stephen LeRoy, who has been arguing for a similar view for a long time, but who does not use the term 'counterfactual' to describe it. I shall also discuss recent work of Judea Pearl, well known for his work on Bayesian nets and causality, econometrician David Hendry and economist/methodologist Kevin Hoover, as well as philosopher of economics, Daniel Hausman. I shall begin with a discussion of some counterfactuals and their uses that I count as genuine, to serve as a contrast to the impostors.

Research for this paper was supported by an AHRB grant, *Causality: Metaphysics and Methods*, and by a grant from the Latsis Foundation. I am grateful to both. Many of the ideas were developed jointly with Julian Reiss (see our paper 'Uncertainty in Econometrics: Evaluating Policy Counterfactuals'); I also want to thank him.

16.2   Genuine counterfactuals

16.2.1   The need for a causal model

Daniel Hausman tells us 'Counterfactual reasoning should permit one to work out the implications of counterfactual suppositions, so as to be prepared in case what one supposes actually happens'.[1] My arguments here will echo Hausman. The counterfactuals that do this for us provide genuine answers to genuine 'What if . . .?' questions; and they play a central role throughout economics. When we consider whether to implement a new policy or try to evaluate whether a trial programme has been successful, we consider a variety of literally intended counterfactual questions: 'What if the policy were put in place?' 'What if the programme had not existed?'

These are just the kinds of questions Heckman considers in his applied work, where he is at pains to point out that the question itself must be carefully formulated. We may for instance want to know what the wages of workers in the population at large would have been had the programme not existed; more commonly we end up asking what the wages of workers in the programme would have been. Or we may want to know what the GDP would have been without the programme. We also need to take care about the contrast class: do we want to know the difference between the results of the programme and those that would have occurred had no alternatives been present or the difference compared to other programmes, real or envisaged?

To evaluate counterfactuals for policy and planning it seems natural to turn to a causal model. Economics is used to producing causal models so this is a good reason to look to economics when we want to evaluate policy counterfactuals. When we want to find out about efficacy, however, there is another method sweeping the social sciences, the method of the randomized controlled trial (RCT), often described as the 'gold standard' for causal inference. This suggests that its results are more certain than any others but that is decidedly not true compared with econometric methods, like more sophisticated versions of the methods of Herbert Simon described in chs. 13 and 14 here. Both are deductive methods, as I point out in ch. 3. So ideally conducted both deliver results as certain as their premises. The trick is of course to conduct them as best possible and in both cases that takes a lot of training as well as a lot of good luck about the subject matter under study – that it will yield to the methods employed to study it.

It is often argued that the RCT uses fewer untested assumptions than an econometric study. I am not so sure about that. It is a little hard to tell how to

---

[1] Hausman (1998), p. 119.

count assumptions. What matters in any given case is how secure we can be about the assumptions that we cannot test – and that will naturally differ from case to case.

This comparison, however, is all about hunting causes. When it comes to using them, which is the topic here, econometric methods have all the advantage because of long practice in building causal models, generally based on a mix of theory, data, intuition and technique. In an RCT, if we are lucky, we find the average difference in effect produced by the treatment in the population sampled. That does not tell us what the overall outcome on the effect in question would be from introducing the treatment in some particular way in an uncontrolled situation, even if we consider introducing it only in the very population sampled. For that we need a causal model. Even less does it tell us about 'side effects' of introducing the treatment, either from the treatment itself or from our way of implementing it. These too are crucial in calculating the costs and benefits of a proposed policy. Or, as Heckman argues, suppose one wants to predict what portion of the population will experience a given degree of improvement. RCTs do not deliver that kind of result. Again we need a causal model.

Building a causal model is not easy. We are all alert to the dangers of taking a model that replicates a pattern observed without policy intervention and supposing that the same pattern will obtain were a given policy implemented. We are also alert to the dangers of taking results that hold in one population, even very well-established results, and building them into a model for a different population. Still, for policy evaluation, a good causal model will serve best. But with the cautions of part I and II in mind, recalling that causal models can model a variety of different kinds of causal relations – we had better be sure it is a causal model of the right kind. I shall take up the question of what is the right kind in section 16.3.4 below.

David Lewis and his followers suppose that we need a model containing the principles by which the system operates (a nomological model) to assess counterfactuals but not a causal model. I do not agree. But it is not this distinction between a Lewis-style merely nomological model and a causal model that I want to discuss. Rather I want to focus on the difference between the causal models that support the counterfactuals we use directly in policy deliberations and those associated with impostor counterfactuals.

For purposes of evaluating a policy counterfactual, besides our causal model we will need to know what changes are envisaged, usually changes under our control. Before that we will need to know what changes are possible. I will turn to this question first, in section 16.2.2 then in 16.2.3 take up the relation between counterfactuals and the changes they presuppose.

16.2.2   What can be changed?

Some people take there to be a universal answer to the question of what can (and should) be changed in assessing counterfactuals. Every separate causal principle can be changed, leaving everything else the same, including all other causal principles, all initial values and all conditional probability distributions of a certain sort. Judea Pearl claims this; as we have seen in part II of this book, so too do James Woodward and Daniel Hausman.

Hausman and Woodward defend this view by maintaining that the equations of a causal model would not represent causal principles if this were not true of them. I have, however, characterized the equations in such a way as to give a different job to them. They are to be functionally correct and to provide a minimal full set of causes on the right-hand side for the quantity represented on the left. The two jobs are different and it would be surprising if they could both be done in one fell swoop, as Hausman and Woodward claim.

Hausman and Woodward object that the jobs cannot be different since the following is true by virtue of the very idea of causation. If a functional relationship between a set of factors (represented by, say, $\{x_j\}$) and a different quantity (say x_e) is functionally correct and the set $\{x_j\}$ is a minimal full set of causes then it must be possible to change this functional relationship, and indeed to stop every one of the x_j from being a cause of x_e, without changing anything else. The x_j would not be causes of x_e were this not true.

I think this claim is mistaken. There are any number of systems whose principles cannot be changed one at a time without either destroying the system or changing it into a system of a different kind. Besides, this assumption does not connect well with other features of causality, described in other accounts, such as probabilistic theories, causal process theories or manipulation accounts.[2]

Pearl has another argument. He says that this assumption is correct because otherwise counterfactuals would be ambiguous. As far as I can tell, the argument must go like this:

---

[2] Hausman (1998) aims to make this connection. But, as his title, *Causal Asymmetries*, suggests, generally what Hausman succeeds in doing is using his claims to obtain causal order. For instance, he shows that, given his claims about the independent variability of causal principles, if b counterfactually depends on a, then a causes b. This is an important result. But to establish it requires the prior assumption that if a and b are counterfactually connected then either a causes b or the reverse or the two have a common cause, plus his own (as opposed for instance to David Lewis's) constraints on the nearness relation for a possible-world semantics for counterfactuals (which I describe below in discussing implementation-neutral counterfactuals). Hausman and Woodward (1999) also claim that the independent variability assumption implies the causal Markov condition. But as I argued in ch. 8, they do not show that the assumption implies the causal Markov condition, which is false, but rather that there are some systems of equations in which both are true and that it is, roughly speaking, 'the same' features of these systems that guarantee both assumptions.

1 Before we can evaluate $c\Box{\rightarrow}e$[3] we must know how c will change, otherwise the counterfactual will be ambiguous.
2 But counterfactuals should not be ambiguous.
3 We can make them unambiguous by assuming that there is a single rule, the same one all the time, about how c will be brought about.
4 The rule that says 'Bring c about by changing the principles that have c as effect to "Set $c = \ldots$."' is such a rule.
5 Therefore we need this rule.
6 But this rule will not be universally applicable unless this kind of change is always possible.
7 Therefore this kind of change must always be possible.

I have written the argument out in detail to make its structure apparent. It is obviously fallacious. It infers from the fact that the rule in question does a needed job that it must be the rule that obtains, which, besides wishful thinking, mistakes a sufficient condition for a necessary one. So I do not think that Pearl's argument will support the conclusion that changes in one principle holding fixed 'everything else' are always possible and indeed are the only possibilities that matter in the evaluation of counterfactuals.

Another similar assumption that is sometimes made is that for the purposes of assessing counterfactuals, changes in the variables of the system are presumed to be brought about by changes in the 'error' terms of the model. But this does not make sense for most interpretations of the error terms. Sometimes these terms are supposed to represent unknown causes omitted from the model. But there is no reason for the unknown causes to be the ones that can change, and when the error terms simply serve to introduce probabilities or to encode measurement inaccuracies, there is not even a quantity there to change. To make sense of the assumption we might instead insist that error terms represent quantities that are 'exogenous' in the sense of 'determined outside the equations that constitute the causal model'. This though will still not guarantee that they can be changed, let alone changed one at a time. Some quantities not determined by the equations of the model will nevertheless be determined by principles outside it, some may not; and some of these outside-the-model principles may be changeable and some may not; and any that are changeable may not be changeable one at a time.

When we consider counterfactuals for the purposes of policy and evaluation, we want change that is really possible, generally without threatening the identity of the system under study. And sometimes it is. What changes are possible and in what combinations, then, is additional information we need for considering policy interventions.

---

[3] I shall throughout use this standard notation for 'If c were the case, e would be the case'.

In the economics literature Kevin Hoover makes this point explicitly.[4] As we have seen in ch. 14 of this book, Hoover distinguishes what he calls parameters from variables. Both vary, but parameters and only parameters can be changed directly by us – any change the value of a variable might undergo will be the result of a change in a parameter. In formulating a causal model, then, it is important to distinguish between the parameters and the variables. We should note, though Hoover himself does not remark on this, that this is not generally the distinction intended between parameters and variables. So we must use care in taking over causal models already formulated that may distinguish parameters and variables in some other way.

16.2.3   What is envisaged to change?

Once we have recorded what things can change, we know what counterfactuals make sense. But to assess the truth-value of any particular counterfactual we will need to know what changes are supposed to happen. Often the exact details matter. For instance, many people feel they would not be opposed to legalizing euthanasia, if only it could be done in a way that would ensure that abuses would not occur.

Sometimes when we consider a policy we have a very definite idea about how it will be implemented. I shall call the related counterfactuals implementation specific. At the other end of the scale, we might have no idea at all; the counterfactuals are implementation neutral. When we evaluate counterfactuals, we had better be clear what exactly we are presuming.

For counterfactuals that are totally implementation specific, we know exactly what we are asking when we ask 'What would happen if . . .?'[5] For others there are a variety of different strategies we might adopt. For one, we can employ the usual devices for dealing with epistemic uncertainty. We might, for instance, assess the probabilities of the various possible methods of implementation and weight the probability of the counterfactual consequent accordingly. In the methodology of economics literature we find another alternative: Stephen LeRoy and Daniel Hausman focus on counterfactuals that would be true regardless of how they were implemented. I begin with LeRoy.

LeRoy's stated concern is with causal ordering among quantities, not with counterfactuals. But, it seems, he equates 'p causes q' with 'if p were to change, q would change as well' – so long as we give the 'right' reading to the counterfactual. It is his proposed reading for the counterfactual that matters here. It may help to present his brief discussion of a stock philosophical example before looking to more formal cases – the case of birth-control pills and thrombosis.

---

[4] Hoover (2001).

[5] Or rather, we know this relative to the factors included in the causal model. Presumably no causal model will be complete, so this remains as a source of ambiguity in our counterfactual claims.

Birth-control pills cause thrombosis; they also prevent pregnancy, which is itself a cause of thrombosis. LeRoy assumes that whether a woman becomes pregnant depends on both her sexual activity and whether she takes pills. Now consider: 'What would happen *vis-à-vis* thrombosis were a particular woman to become pregnant?' That, LeRoy, points out, is ambiguous – it depends on whether the change in pregnancy comes about because of a change in pill-taking or because of a change in sexual activity.

In his formal characterization LeRoy treats systems of linear deterministic equations. We may take these to be very sparse causal models. They are what in economics are called 'reduced form equations': 'In current usage an economic model is a map from a space of exogenous variables – agents' characteristics and resource endowments, for example – to a space of endogenous variables – prices and allocations.'[6] The equations are expected to be functionally correct, but not to represent the causal relations among the variables, with one exception. Variables designated as 'exogenous' are supposed not to be caused by any of the remaining (endogenous) variables. Since they are functionally related to the endogenous variables, we may assume that either they are causes of some of the endogenous variables or are correlated with such causes. For LeRoy's purposes I think we must suppose they are causes.

In the context of our discussion here, with Hoover in mind, we should note one further assumption that LeRoy makes. The possible sources of change in an endogenous variable are exactly the members of the minimal set of exogenous variables that, according to the economic model used to evaluate the counterfactuals, will fix the value of the endogenous variable. LeRoy considers a familiar supply and demand model:

$$q_s = \alpha_s + \alpha_{sp}p + \alpha_{sw}w \tag{1}$$
$$q_d = \alpha_d + \alpha_{dp}p + \alpha_{di}i$$
$$q_s = q_d = q$$

where p is price; q, quantity; w, weather; i, income. LeRoy asks what the effect of a change in price would be on the equilibrium quantity. By the conventions just described, a change in price can come about through changes in weather, income or both, and nothing else. But, LeRoy, notes, 'any of an infinite number of pairs of shifts in the exogenous variables "weather" and "income" could have caused the assumed changes in price, and these map on to different values of q'.[7] Thus the question has no definite answer – it all depends on how the change in p is brought about.

LeRoy contrasts this model with a different one:

$$q_s = \alpha_s + \alpha_{sw}w + \alpha_{sf}f \tag{2}$$
$$q_p = \alpha_p + \alpha_{dp}p + \alpha_{di}i$$
$$q_s = q_d = q$$

[6] LeRoy, 2003, p. 1.   [7] LeRoy, 2003, p. 6.

where f is fertilizer. Here fertilizer and weather can change the equilibrium quantity, and no matter how they do so, the change in price will be the same. In this case Leroy is content that the counterfactual, 'If q were to change from Q to $Q + \Delta$,[8] p would change from $P = (Q - \alpha_p - \alpha_{di}I)/\alpha_{dp}$ to $P = (Q + \Delta - \alpha_p - \alpha_{di}I)/\alpha_{dp}$' is unambiguous (and true). The lesson he draws is the following (where I substitute counterfactual language for his causal language): '[Counterfactual] statements involving endogenous variables as [antecedents] are ambiguous except when all the interventions consistent with a given change in the [antecedent] map onto the same change in the [consequent].'[9] I think the statement as it stands is too strong. Some counterfactuals are, after all, either implicitly or explicitly implementation specific. What LeRoy offers is a semantics for counterfactuals that are, either implicitly or explicitly, implementation neutral. In this case the consequent should obtain no matter what possible change occurs to bring the antecedent about.

Daniel Hausman seems to have distinguished between implementation-specific and implementation-neutral counterfactuals, too, as I do here, though I do not think he explicitly says so. He considers an example in which engineers designing a nuclear power plant ask, 'What would happen if the steam pipe were to burst?'[10] The answer, he argues, depends on how it will burst. 'Responsible engineers', he argues, must look to the origins of the burst 'when the consequences of the pipe's bursting depend on what caused it to burst.'[11]

On the other hand, when Hausman turns to providing some constraints that a possible-world semantics for counterfactuals must satisfy, he seems to be concerned with implementation-neutral counterfactuals. The results are similar to LeRoy's. Any semantics that satisfies Hausman's constraints should give the same result as LeRoy's prescription when restricted to counterfactuals evaluated via what LeRoy calls an 'economic model'. The Hausman constraint on the similarity relation between possible worlds that matters to our discussion here is

SIM2. *It doesn't matter which cause is responsible.* For any event b, if a and c are any two causes of b that are causally and counterfactually independent of one another, there will be non-b possible worlds in which a does not occur and c does occur that are just as close to the actual world as are any non-b possible worlds with a and without c, and there will be non-b possible worlds without a and with c that are just as close to the actual world as are any non-b possible worlds without both a and c.[12]

Look back at LeRoy's model (1) for illustration, where weather and income are the causes by which either price or quantity can change. It is easiest to see the results if we first solve for p and q:

$$q = (\alpha_{dp}\alpha_s - \alpha_{sp}\alpha_d + \alpha_{dp}\alpha_{sw}w - \alpha_{sp}\alpha_{di}i)/(\alpha_{dp} - \alpha_{sp})$$
$$p = (\alpha_s - \alpha_d + \alpha_{sw}w - \alpha_{di}i)/(\alpha_{dp} - \alpha_{sp})$$

---

[8] I shall follow LeRoy's convention throughout and use lower-case letters for variables and upper case for their values.
[9] LeRoy (2003), p. 6.   [10] Hausman (1998), p. 122.
[11] Hausman (1998), p. 122.   [12] Hausman (1998), p. 133.

If p changes by ΔP with w fixed, then i must have changed by $\Delta P(\alpha_{sp} - \alpha_{dp})/\alpha_{di}$ and so q will change by $\Delta Q = \alpha_{sp}\Delta P_i$. If on the other hand i is fixed, then w must have changed by $\Delta W = \Delta P(\alpha_{dp} - \alpha_{sp})/\alpha_{sw}$ and so $\Delta Q = \alpha_{dp}\Delta P$. Now we can bring in SIM2). If q changes (q is here the analogue of b in SIM2) some world in which w (the analogue of a) changes will be just as close as any world in which i (the analogue of c) changes. But the world in which w changes and i stays fixed and the world in which i changes and w stays fixed have different values for the change in q. Yet they are equally close. So the truth values of counterfactual claims about what would happen to q were p to change by ΔP are undefined.

So we may have counterfactuals that are implementation specific; we may have ones that assume some one or another of a range of possible implementations; and we may have implementation-neutral ones where we wish to find out what would happen no matter how the change in the antecedent is brought about. For thinking about policy we had better know which kind of counterfactual we are asserting and ensure that our semantics is appropriate to it.

16.3   Impostor counterfactuals

The kinds of 'What if . . .?' questions asked in planning and evaluating are in sharp contrast with a different kind of 'counterfactual' that occupies economists as well – the impostor counterfactuals. Like the counterfactuals I have so far been discussing these too are evaluated relative to a causal model. But they are not used directly in planning and evaluation. Rather they are used to define certain causal concepts. For Heckman the relevant concept is causal effect; for LeRoy, causal order. I shall discuss LeRoy first.

16.3.1   LeRoy

I have urged that in order to assess policy counterfactuals, the best idea is to have a causal model. Recall that LeRoy begins with a sparse causal model: a reduced form equation that links the endogenous variables to a set of exogenous variables, where he supposes that no exogenous variables are caused by any endogenous ones and that the exogenous variables completely determine the values of the endogenous variables.[13] The task is to say something about the causal order of the endogenous variables and, I take it, about the strength of influence of one on another. Let Z_j be the minimal set of exogenous variables

---

[13] Note that the reduced form equation need not be a causal function in the sense that I shall introduce from Heckman, since LeRoy allows that the external variables may not be variation free, though he thinks it would be odd if they were not.

that determine x_j, define Z_{ji} as $Z_j - Z_i$ and $\overline{Z}_{ji}$ as the vector of elements in Z_{ji}. Then x_c causes x_e if and only if there is a (scalar) γ_{ec} and a (vector) δ_{ec} such that

$$x_e = \gamma_{ec}x_c + \delta_{ec}\overline{Z}_{ec}$$

This means that x_e is determined completely by x_c plus a set of exogenous variables that do not participate in determining x_c; that is, there is no z that both helps fix the first term in the above equation and also helps fix the second.

What what-if question does γ_{ec} answer? It answers an implementation-neutral counterfactual: by how much would x_e change were x_c to change by a given amount, no matter how the change in x_c is brought about? This is often an important question to be able to answer; it may also be important to know for the system we are dealing with that it has no answer: there is nothing general, or implementation neutral, that we can say; how much the effect changes cannot be calculated without knowing what the method of implementation will be.

There are two points I would like to make about LeRoy's approach. First I admit that these counterfactuals are in no way 'impostors' – they ask genuine what-if question whose answers we frequently need to know. Nevertheless they are severely restricted in their range of application. For vast numbers of systems the answer to LeRoy's counterfactual question will be that it has no answer: there is no implementation-neutral change that would occur in the effect consequent on a change in the cause.[14]

Second, LeRoy's definition answers one very special kind of causal question – it asks about how much, if one factor changes in any way whatsoever, a second factor will change. But it does not answer the question of how much one factor contributes to another. For a simple example where the two questions have different answers consider a system governed by the following two causal laws:[15]

$$q_c = \alpha_{qz}z \qquad\qquad\qquad\qquad\qquad\qquad\qquad (\text{CM1})$$
$$p_c = \alpha_{pz}z$$

Compare this with a system governed by different laws

$$p_c = \alpha_{pz}z \qquad\qquad\qquad\qquad\qquad\qquad\qquad (\text{CM2})$$
$$q_c = \alpha_{qp}p$$

It should at least in principle be possible for two such systems to exist. The two systems have different causal structures and different answers to the question, 'How much does p contribute causally to q?' In the second system the

---

[14] It should be noted that often implementation-specific counterfactuals will have no truth value either. This happens whenever changes in the variable in the antecedent plus the implementation variable can occur in more than one way with different effects on the consequent.

[15] I use here the same notation for causal laws as I have throughout this book.

answer is given by α_{qp}. In the first the answer is 'nothing'. Yet in cases where $\alpha_{qz} = \alpha_{qp}\alpha_{pz}$ there will be exactly the same answer to LeRoy's counterfactual question: if p were to change by Δp, no matter how it does so q would change by $\alpha_{qz}\Delta p = \alpha_{qp}\alpha_{pz}\Delta p$.

As I argued in part I, we have a large variety of causal concepts, applicable to a variety of different kinds of systems in different situations. So, too, there is a large variety of different kinds of causal and counterfactual questions we can ask, many of which only make sense in particular kinds of systems in particular circumstances. LeRoy asks a specific, explicitly articulated counterfactual question, and I take it that that is all to the good. We must be careful, however, not to be misled by his own use of the language of 'causal order' to suppose it tells us whether and how much one quantity causally contributes to another.

16.3.2   Heckman

Heckman also uses counterfactuals to answer what he labels as causal questions, and he is very careful to insist that we ask a clear, well-articulated counterfactual question. Generally these are implementation-specific questions. For some of these – and these are the ones I worry about – as with LeRoy, the question has an answer only in certain restricted systems – essentially, as I shall explain, in Galilean-style experiments. As far as I can see, the primary interest in these counterfactuals is that they serve as a tool for answering a non-counterfactual question, a question about causal contributions. But questions about causal contributions can be asked – and answered – for situations that are not Galilean experiments, where the counterfactuals Heckman introduces do not make sense. This is why I say that they are impostors. They seem to be the issue of interest; they are certainly the topic. But in fact they are only a tool for answering a different question – a causal question – and at that, for answering that question only in very restricted kinds of systems, kinds that are not generally the ones of concern.

Before we turn to Heckman it may be helpful to begin with work that will be more familiar to philosophers, from the book *Causality* by Judea Pearl. Pearl gives a precise and detailed semantics for counterfactuals. But what is the semantics a semantics of? What kinds of counterfactuals will it treat, used in what kinds of contexts? Since Pearl introduces them without comment we might think that he had in mind natural language counterfactuals. But he presents only a single semantics with no context dependence, which does not fit with natural language usage.

Worse, the particular semantics Pearl develops is unsuited to a host of natural language uses of counterfactuals, especially those for planning and evaluation of the kind I have been discussing. That is because of the special way in which he imagines that the counterfactual antecedent will be brought about: by a precise incision that changes exactly the counterfactual antecedent and nothing else (except what follows causally from just that difference). But when we consider implementing a policy, this is not at all the question we need to ask. For policy and evaluation we generally want to know what would happen were the policy really set in place. And whatever we know about how it might be put in place, the one thing we can usually be sure of is that it will not be by a precise incision of the kind Pearl assumes.

Consider for example Pearl's axiom of composition, which he proves to hold in all causal models – given his characterization of a causal model and his semantics for counterfactuals. This axiom states that 'if we force a variable (W) to a value w that it would have had without our intervention, then the intervention will have no effect on other variables in the system'.[16] This axiom is reasonable if we envisage interventions that bring about the antecedent of the counterfactual in as minimal a way as possible. But it is clearly violated in a great many realistic cases. Often we have no idea whether the antecedent will in fact obtain or not, and this is true even if we allow that the governing principles are deterministic. We implement a policy to ensure that it will obtain – and the policy may affect a host of changes in other variables in the system, some envisaged and some not.

We should note that the same problem arises for Lewis-style semantics. If the antecedent of a counterfactual obtains, then our world, with things as they actually happen in it, is the nearest possible world for evaluating the truth value of the counterfactual. There is no room then for anything to change as a result of the antecedent being implemented.[17]

Heckman, unlike Pearl and Lewis, is keen that causal models model how change is brought about. So in defining causal efficacy he does not adopt Pearl's semantics in which laws are changed *deus ex machina*. But he does adopt a device that I think is similar. Pearl limits his causal definitions to systems in which the principles responsible for a given factor with all their causes, can be changed to produce any required value for that factor, without changing any other principles or other 'initial' values. Heckman limits his definitions to causal principles in which the causes are variation free. This means that if only the system runs 'long enough', the effect (intended as the antecedent of the counterfactual) will naturally take any required value, while the remaining causes, all other principles, and all other initial values stay the same. The counterfactual change in an antecedent with 'everything else' the same will 'eventually' be factual. Heckman stresses, thus, that what matters for his definitions is natural variability within the system, not changes in the principles under which it operates.

---

[16] Pearl (2000), p. 229.
[17] For a longer discussion of Pearl and Lewis see Reiss and Cartwright (2003).

Heckman begins his treatment with causal functions. These govern very special kinds of causal system, systems that mimic experiments:

Causal functions are . . . derived from conceptual experiments where exogenously specified generating variables are varied. . . . The specification of these hypothetical variations is a crucial part of model specification and lies at the heart of any rigorous definition of causality.[18]

Heckman tells us three things about causal functions: (1) They 'describe how each possible vector of generating variables is mapped into a resulting outcome', where the generating variables 'completely determine' the outcome.[19] (2) They 'derive from' – or better, I think, 'describe' – conceptual experiments. (3) Touching on questions of realism and of model choice, models involving causal functions are always underdetermined by evidence; hence, as Heckman sees it, causality is just 'in the mind' since the models relative to which it is defined are just in the mind. From this I take it that causal functions represent (a probably proper subset of) the causal principles under which these special experiment-like systems operate, where the right-hand-side variables – the ones Heckman calls the 'generating variables' – form a minimal complete set of causes of the quantity represented on the left[20] and where each cause can vary independently of the others.

Imagine that the causal function for an outcome y is given by

$$y = g(x_1, \ldots, x_n)$$

We can now define the causal or counterfactual effect of x_j on y fixing the remaining factors in the causal function (Heckman seems to use the terms 'causal effect' and 'counterfactual effect' interchangeably):

(causal effect of x_j on y)

$$[\Delta y / \Delta x_j] =_{df} g(x_1, \ldots, x_j', \ldots x_n) - g(x_1, \ldots, x_j'', \ldots x_n)$$

where

$$\Delta x_j = x_j' - x_j''$$

As Heckman insists, in order for this definition 'to be meaningful requires that the x_j can be independently varied when the other variables are fixed so that there are no functional restrictions connecting the arguments . . . it is thus required that these variables be variation-free'.[21] I shall call the counterfactual

---

[18] Heckman (2001), p. 14.   [19] Heckman (2001), p. 12.
[20] Or, keeping in mind Heckman's view that causality is only relative to a model, the right-hand-side variables record what the model designates as causes.
[21] Heckman (2001), p. 18.

effect as thus defined a Galilean counterfactual since, as I remarked, it is just the kind of effect we look for in a Galilean experiment.

I should note that Heckman himself treats of double counterfactuals since the outcome variables he discusses are often themselves counterfactuals: y_0 is the value a given quantity would take were a specified 'treatment' to occur; y_1, the value it would take were the treatment not to occur. These values, he supposes, are fixed by deterministic causal functions. Relative to these causal functions we can then ask about the causal efficacy of a certain quantity – including the treatment itself – on the counterfactual quantities y_0 and y_1. So we can consider, for example, what difference a change in social security regulations would have on the amount of savings that would obtain if there were a tax cut versus the difference the change would make were there no tax cut. I will not be concerned with these double-barrelled counterfactuals here. They do not appear in Heckman's discussion of the supply and demand equations, which will suffice as illustrations of my central point.

Heckman considers simultaneous supply and demand equations. For simplicity we can look at the specific equations that we have already considered above, where I have added the additional equilibrium constraint on price:

$$q_s = \alpha_s + \alpha_{sp}p_s + \alpha_{sw}w \qquad (1')$$
$$q_d = \alpha_d + \alpha_{dp}p_d + \alpha_{di}i$$
$$q_s = q_d = q$$
$$p_s = p_d = p$$

Heckman points out that these equations do not fit Pearl's scheme since they are not recursive and hence Pearl's method for assessing counterfactuals will not apply. This fits with familiar remarks about these kinds of systems: p and q are determined jointly by exogenous factors. It seems then that it makes no sense to ask about how much a change in p will affect a change in q. To the contrary, Heckman points out: We can still assess causal efficacy using his definition – so long as certain 'exclusion' conditions are met.

Suppose we want to assess the causal/counterfactual effect of demand price on quantity demanded. We first look to the reduced form equations

$$q = (z_d, z_s)$$
$$p = (z_d, z_s)$$

where z_d is the vector of exogenous variables in the demand equations and z_s, those in the supply equations. In LeRoy's equations $(1')$, $z_d = i$ and $z_s = w$. Heckman takes these to be causal functions, otherwise the causal model has not properly specified the 'exogenous' variables. That means that the exogenous variables are 'generating variables' for p and q and that they are variation free. Now the task is easy:

Assuming that some components of $[z_d]$ do not appear in $[z_s]$, that some components of $[z_s]$ do not appear in $[z_d]$, and that those components have a non-zero impact on price, one can use the variation in the excluded variables to vary $[p_d$ or p_s in the reduced form equations] while holding the other arguments of those equations fixed.[22]

The result (using the equality of p_d and p_s and of q_d and q_s) is

$$\partial q_d / \partial p_d = (\partial q / \partial z_s(e)) / (\partial p / \partial z_s(e))$$

where $z_s(e)$ is a variable in z_s that is excluded from z_d and that, as he puts it, 'has an impact on' p_d. In (I') this job can be done by w; the causal effect thus calculated of p_d on q_d is α_{dp}.

Notice how much causality is involved here. By definition we are supposed to be evaluating the change in q_d holding fixed all the factors in a causal function for q_d except p_d. What we actually do is hold fixed z_d while z_s varies. Though this does not fit the definition exactly, presumably this is okay because z_s is a cause of p_d that can produce variations in p_d while z_d is fixed; and z_d being fixed matters because z_d constitutes, along with p_d, a minimal full set of causes of q_d. So when the exclusion condition is satisfied, the demand equation is a causal function and the counterfactual definition of causal effect is meaningful.

Now consider a slightly altered set of equations:

$$q_s = \alpha_s + \alpha_{sp} p_s + \alpha_{sw} w + \alpha_{si} i \qquad (1'')$$
$$q_d = \alpha_d + \alpha_{dp} p_d + \alpha_{di} i + \alpha_{dw} w$$
$$q_s = q_d = q$$
$$p_s = p_d = p$$

Now the demand equation cannot be treated as a causal function and the question of the causal effect of demand price on quantity demanded is meaningless. This is true despite the fact that α_{dp} still appears in the equation and it still represents something – something much the same one would suppose – about the bearing of p_d on q_d. The intermediate case seems even stranger. Imagine that $\alpha_{sw} = 0$. Now α_{sp} measures a counterfactual effect but α_{dp} does not.

16.3.3   Cartwright

I have an alternative. But I should note that I have a stake in this discussion since I have been stressing the importance of independent variability for over fifteen years; I just think it plays a different role from the one Heckman (and Pearl and Hausman and Woodward) ascribe to it.

I begin with causal principles. At this level of discussion I myself am a realist about the principles of our causal models: they are correct if and only if they approximate well enough to the causal laws that govern the operation of

---

[22] Heckman (2001), p. 36.

the system in question. Heckman, it seems, is not a realist. But that does not matter here since he himself has introduced the notion of a causal function. A causal principle is just like a causal function but without the restriction that the causes (or 'generating variables') are variation free. I shall restrict attention to linear deterministic systems, as defined in ch. 10 of this book. Then, for a given causal system, the contribution a cause x_c makes to an effect x_e is just the coefficient of x_c in any causal principle for x_e in the system.[23] It is easy to show for linear deterministic systems that where Heckman's measure for the causal/counterfactual effect of x_c on x_e applies, it will have the same value as the contribution x_c makes to x_e.

Given this characterization we see that the contribution of p_d to q_d is the same in (1') and (1''). What is different is that in (1') we have a particular way to find out about it that is not available in (1''). (1') is what in ch. 10 I call an epistemically convenient system. It is a system in which we can find out what a cause, x_c, contributes to an effect, x_e, in one particular simple way: hold fixed all the other contributions that add up to make the effect the size it is; then vary the cause and see how much x_e varies. Any difference has to be exactly the contribution that x_c adds. This does not mean, however, that for systems where this independent variation is not possible, all is lost. There are hosts of other legitimate ways of defending claims about the size of causal contributions that apply both in systems with independent variation and in ones without.[24]

16.3.4 Hendry and Hoover

To set policy in an informed way, a causal model is required. What is a causal model? Ideally it is a model that contains enough information to calculate all the consequences that might matter about all the possible changes envisaged. This characterization is exceedingly thin and gives no help about how to set about constructing a causal model. But it is probably the most specific characterization that can be given without focusing on particular kinds of situation. That is in keeping with the claim that there are a great variety of kinds of causal relations embedded in a great variety of kinds of causal systems as well as a variety of causal questions that can be asked.

---

[23] Recall that the discussion here is limited to linear systems; the concept of a causal contribution is more complex in non-linear systems. Also note that this supposes that all principles in the model with x_c on the right-hand-side and x_e on the left will have the same coefficient. This will be the case given a proper statement of 'transitivity' and the definitions for the form of causal principles sketched in ch. 10.

[24] For further discussion see Cartwright (1989). It should be admitted of course that once the causes need not be variation free, the simple operational way of defining causal contribution in a way analogous to Heckman's definition of causal/counterfactual effect is not available. But, as we know, there are compelling arguments in the philosophical literature to establish that demanding operational definitions is both too strong and too weak a requirement – it lets in concepts that do not make sense and does not provide a proper understanding of those that do.

Consider the two versions of Simon discussed in chs. 13 and 14. The first, mine, is like Judea Pearl's and it shares the same vices and virtues; in particular it has two special drawbacks when it comes to evaluating genuine 'What if . . .?' counterfactuals. First, it is geared simultaneously to hunting and to using causes. The hunting part constrains it enormously. The systems studied are identifiable and have the characteristics that ensure that the equations identified are causally correct. That limits their scope dramatically. This is analogous to the problem raised with Heckman's notion of causal effect. Second, an additional recipe is required for how to read off the truth values of counterfactuals. Simon supplies this in 1953.[25] Essentially changes propagate 'down' through the triangular array of equations in the order in which they can be solved using the exogenous variables. Pearl provides a more elaborate semantics for more complicated forms of counterfactual questions in *Causality*,[26] but clearly Pearl's and Simon's should agree where both make verdicts about the same question.

These two drawbacks do not affect the kind of causal relations that Hoover studies, which I called 'strategy relations'. His relations are not restricted to systems that can be identified from probabilities; they allow for other methods of warrant.[27] And they wear the semantic for 'What if . . .?' counterfactuals on their sleeve. All and only the things we can change are represented by parameters and the strategy relations tell what values the variables take given values for the parameters.

David Hendry maintains that causal relations must be superexogenous.[28] Like Simon, as I represent him, Hendry seems to restrict his notion of causality to systems in which causes can be hunted by straightforward econometric means – the 'causes' are exogenous with respect to the parameters of interest; like Hoover, the use for licensing counterfactuals is immediate – that comes from the demands for superexogeneity.

Consider as a simple example a vector Y of outcome variables that we consider affecting via changes in the parameters, γ, of the probability of x. Expressing the joint probability as $P(Y/x, \beta)P(x, \gamma)$, x is weakly exogenous to Y if the parameters γ of the marginal distribution have no cross-restraints with the parameters β of the conditional distribution. This means that the marginal distribution can be ignored without loss of information in estimating the conditional distribution. This is useful since under these conditions the joint distribution can be estimated by separate estimation of the conditional and marginal distributions.

Since the parameters have no cross-restraints it may look as if we can change γ without affecting β. That is not correct. These facts about the relation of

---

[25] A good analysis of what Simon does there can be found in Fennell (2005a).
[26] Pearl (2000), p. 525.
[27] Though, as I argued, they are highly restricted by the demand that the variables be determined by the parameters since parameters by definition are quantities over which we have direct control.
[28] See Hendry (2004), Engle, Hendry and Richard (1983) and ch. 4 in this book.

the parameters are facts that hold for the given distribution P. To change the marginal distribution in any way is to change the distribution P to some new one, P'. There is nothing in the facts reported about P that has bearing on what happens in P'. Exogeneity in P provides no guarantee that the parameters for the joint and marginal distributions will factor in P' with γ in the marginal distribution alone or that the conditional distribution will stay fixed. To demand that they do so is to demand the superexogeneity of x for Y with respect to the parameters γ.

Consider a simple two equation system such as

$$x = \gamma + u_1$$
$$Y = \beta x + u_2$$

where γ and β are parameters that have no cross restraints (are variation free) and u_1 and u_2 are normalized independent error terms. In this system, weak exogeneity holds since γ and β are variation free. However, suppose that the parameters we actually intervene on are α_1 and α_2 where

$$(\beta, \gamma) = f(\alpha_1, \alpha_2)$$

and f is invertible (so that γ and β are variation free if the α's are).

Now suppose we propose to intervene on x in order to affect Y. Ideally we should like to fix γ. But we can only do so via intervening on α_1 or α_2, and intervening to change γ by acting on α_1 and α_2 also changes β. So changing the marginal distribution $P(x, \gamma)$ by changing β also changes the conditional distribution $P(Y|x, \gamma)$ despite the fact that γ and β are variation free. So, one needs to strengthen the weak exogeneity requirement so that one can intervene to change parameters of either the marginal distribution or conditional distribution without affecting the parameters of the other distribution. This is to require superexogeneity.

Turning now to counterfactuals. Given superexogeneity it is possible to read off from the conditional distribution how the probability of the outcomes of interest will change if changes in the distribution of x are implemented in the ways envisaged (i.e. by changing γ). No intermediate semantics is required. Are these genuine 'What if . . .?' counterfactuals or mere impostors? Since these are not technical terms but merely a way of making a point about the difference between causal discovery – entering the language of causality – and causal use – exiting that language, it should not be surprising that the answer is that they are 'somewhere in between'. What Hendry calls causal relations do tell us what would happen if what we are actually thinking of doing in the way we thinking of doing it were to happen. But they do so only for systems where superexogeneity is ensured, which, though perhaps not as demanding as the conditions for a Galilean experiment, is highly restrictive when it comes to real life.

16.4   Epistemic convenience versus external validity

I began my discussion with reference to impostor counterfactuals. There is a sense in which the Galilean counterfactual questions that Heckman asks are genuine. If we are talking about the right kinds of system – epistemically convenient ones – they ask genuine implementation-specific what-if questions. But there are two problems. First, few systems we confront are epistemically convenient. The vast majority are not. For these, the particular measures Heckman introduces under the title 'causal (or counterfactual) effect' are irrelevant.

Second, even if we are studying an epistemically convenient system there is a puzzle about why we should wish to ask just these implementation-specific questions. If we were thinking of setting policy or evaluating the success of some programme in the system, then these, with their very special method of implementation, might be relevant sometimes. But there is no necessity to implement policies in the single way highlighted in the Galilean counterfactual; generally we would want to consider a variety of different methods of implementation and frequently to assess implementation-neutral counterfactuals as well. Even in epistemically convenient systems, the Galilean counterfactuals that Heckman studies often have no privileged role.

There are two familiar enterprises where they do have a special role. The first is in trying to determine if, and to what degree, one factor contributes causally to another. In an epistemically convenient system we can ask Galilean-type counterfactual questions; and the answers we obtain will double as measures of causal contributions. They are a tool for finding out answers to our causal questions. But note that they are only a tool for finding out about causes in our special epistemically convenient systems. For other systems we cannot even ask these counterfactual questions, let alone let the answers to them supply our causal answers as well.

The other is in Heckman's own field, evaluation. In setting up new programmes, we might try to set them up in such a way that the causal contribution they make to the result can be readily disentangled from the contribution of other factors. One particular concern is with other factors that might both contribute to the effect independently of the programme and also make it more likely that an individual entered (or failed to enter) the programme. If we can arrange the set-up of our programme so that it is epistemically convenient, then again we can answer Galilean counterfactual questions – 'What difference would there be in outcome with the programme present versus the programme absent, holding fixed all other contributions to the outcome?' And again these counterfactual questions will tell us the contribution the programme makes, since in these circumstances the difference in outcome between when the programme is present and when it is absent must be exactly the contribution the programme makes. So we can use information about Galilean counterfactuals to learn about the causal contributions of the programme we set up. Still, all we learn is about that programme in those special epistemically convenient circumstances.

In either case, whether it be experimental systems or programme set-ups that we engineer to make the measurement of causal contributions easy, we need to ask why we should be interested in causal contributions in these very special – and rare – kinds of system. The answer is clear. Generally we want this information because it will tell us something about causal contributions in other systems as well. But we confront here the familiar problem of internal and external validity. In an epistemically convenient linear system, using counterfactual differences as a measure of causal contributions is provably valid. Internal to the situation this method is bound to give us correct results about the question of interest. But nothing said in this discussion bears on external validity: when will the results that we can be sure are correct in a convenient system hold elsewhere? Sometimes this issue is discussed in the economics methodology literature under the heading 'invariance'. This is often with something like equation set (l′) in mind. Here we can find out the causal contribution, α_{dp}, of p_d to q_d by calculating the difference in Galilean counterfactuals as p_d changes via w holding fixed i. Then we might imagine that everything already in place about the causal principle for q_d would stay the same even if weather became an influence on quantity demanded. Thus we suppose that the second equation can be replaced with

$$q_d = \alpha_d + \alpha_{dp} p_d + \alpha_{di} i + \alpha_{dw} w$$

We then say that the equation for q_d remains invariant as α_{dw} changes from zero to non-zero, or possibly we suppose it invariant over any range of values for α_{dw}. This though is only one kind of assumption we might make about the use to which we can put the information we learn about the causal contribution that one factor makes to another.

There are two points that matter to my argument here. The first is that assumptions about where this information can be put to use are not justified by anything we have discussed so far, and in particular not by any information about counterfactuals of the kinds I have explored. Showing that results on causal contributions have external validity – and how far and of what kind – requires a different methodology altogether.

Second, when we export the information gleaned from Galilean counterfactuals in epistemically convenient systems elsewhere, it is not as information about counterfactuals but rather as information about causal contributions. In most systems to which we will carry our results, Galilean counterfactual questions do not even make sense. This supports my claim that both as counterfactuals and as causal surrogates, Galilean counterfactuals are impostors. They do not carry over as counterfactuals to non-epistemically convenient systems; and in epistemically convenient ones they are usually of interest, not on their own as genuine what-if hypotheses but only as tools for measuring causal contributions. Even then the results about causal contributions are of use outside the highly restricted systems in which they are established only if specific assumptions about the external validity of the results are warranted.

This issue links directly with the discussion of RCTs in section 16.2.1 of this chapter and in ch. 3 of this book. Consider a linear three-variable example for simplicity. Suppose smoking (x_1) and exercise (x_2) are a complete set of causes for degree of heart health (y), where smoking and exercising can take values 0 and 1. So the causal function is $y = ax_1 + bx_2$ and the causal effect of smoking on heart health is a. Imagine we do not know the full 'causal function' for the outcome y so we use an RCT to try to learn the size of a, for illustration a very ideal RCT where everyone in the treatment group smokes and no one in the control group smokes and where, as hoped for from the randomization, the probability of exercise is the same in both. Then, using subscripts T and C for the treatment and control groups, and noting that the mean value of x_2 should be the same in both groups

$$\langle y \rangle_T - \langle y \rangle_C = a + b(\langle x_2 \rangle_T - \langle x_2 \rangle_C) = a$$

So the causal effect, a, shows up as the difference in the mean outcomes of the two groups.

What does this tell us about what would happen in real life were we to induce everyone to stop smoking? Suppose we stick with the simplest case where we are concerned only with the population from which the sample for the experiment was drawn and we suppose our policies do not affect the underlying causal function. In the experiment we create an artificial situation where smoking and exercise vary separately from one another. But this is unlikely to be true outside the experiment. Free from the experimental gaze, people who would otherwise have exercised might not do so; others may start to do so; and being induced to stop smoking in whatever way the policy employs may itself lead people to change their exercise habits. The problems are compounded if we imagine trying to carry the results to new populations or suppose that new kinds of causes occur. In this case we have no assurance so far that even the underlying causal function will stay the same.

This is well known. I repeat it to underline that the size of the causal effect is not the same as the counterfactual difference for real counterfactuals outside experiments. To calculate that we need a causal model of what happens in the real case with the kind of policy implementations envisaged. In building a causal model the information that the causal difference in an epistemically convenient system somehow related to ours, or in a well-conducted RCT, is a can be of help. But to import this into our new model requires a number of strong assumptions well beyond those required to determine that the causal effect has a certain size in the epistemically convenient system or in the RCT.

The same kinds of concern have been raised about my views on capacities. One of the two central themes of this book is causal diversity: there are a great variety of different kinds of causal concepts. My recognition of this began with a distinction between causal-law claims and capacity claims. One chief kind of example of causal-law claims are the equations of a linear deterministic system. I took it that these describe the principles that govern the production relations[29] of a certain kind of structure or institution, like a toaster of a certain make or the UK economy in 2003. Capacity claims describe facts about causes that hold more widely. We can learn about them in one structure or setting and use what we learn there to construct new causal models for new situations.

I had in mind the long-standing worries in economics – by Mill,[30] Frisch,[31] Haavelmo,[32] Lucas[33] and Friedman[34] among others – that economic parameters may not be stable as the causal principles for a situation change. In our little smoking/exercise/health example, we may lessen the bad effects of smoking on heart health, which are measured by a, by regulating the contents of tobacco; but that intervention may also change how much exercise can affect heart health, measured by b. In that case b does not measure a stable capacity that can be relied on in the building of a causal model for the new situation.[35]

In *Nature's Capacities and their Measurement* I argued that empiricists need not shun capacities because they are not measurable. We can measure them in Galilean experiments or using standard econometric techniques – and we do so all the time. But, it has been objected,[36] the measurements measure the strength of a capacity given that it is a capacity we are measuring. They do not show that it is a capacity. That is indeed the case. It is the reason for stressing the peculiarity of the situation of the Galilean experiment or the epistemically convenient system. These are special rare kinds of situation. What happens in them would generally be of little consequence for practice unless we have good reason to suppose that what happens there is characteristic of what happens elsewhere. But that is a separate matter needing its own independent and different kinds

---

[29] Of course, as I note in my discussion in chs. 2, 7 and 14, I no longer think that there is one single kind of causal relation that equations of this form might describe.

[30] Mill obviously did not talk about parameters. But he did stress that the principles observed to describe a system correctly cannot be relied on to continue to do so in the future because the background arrangement of causes giving rise to any observed pattern is likely to shift unpredictably. See Mill (1836).

[31] Frisch (1938).   [32] Haavelmo (1944).   [33] Lucas (1976).   [34] Friedman (1953).

[35] Notice that this kind of invariance, which I mark with the term 'capacity', is different from any of the invariances discussed in chs. 8–10 in this book. It is also different from another kind of invariance studied by James Woodward (2003) and Sandra Mitchell (2003), where we ask for a given set of principles how widely they apply.

[36] Morrison (1995).

of arguments.[37] The Galilean experiment and the epistemically convenient system have just the right structure to allow us to figure out what is happening within them. But nothing in that structure argues that the results can be carried elsewhere.

16.5   Causal decision-theory

As another illustration of the conflation of Galilean counterfactuals with more realistic implementation-specific ones, consider causal decision-theory. Various versions of causal decision-theory made the same mistake I am pointing to, but in reverse. The aim was to evaluate genuine counterfactuals but we ended up with a measure that measured the causal contribution of a factor and not the counterfactual effects of the factor being implemented. Consider a very simple case.

Given my fear of lung cancer, should I quit smoking? Presumably the answer is 'yes' if the expected utility if I were to quit is greater than if I were to continue; or

Counterfactual decision formula:

$$P(S^{\square}{\rightarrow}L)U(S\&L) + P(S^{\square}{\rightarrow}\neg L)U(S\&\neg L) < P(\neg S^{\square}{\rightarrow}L)U(\neg S\&L)$$
$$+ P(\neg S^{\square}{\rightarrow}\neg L)U(\neg S\&\neg L)$$

where $S = $ I smoke, $L = $ I get lung cancer, $U(X) = $ utility of X, and where I shall assume the probabilities are personal probabilities read off from the population probabilities.

Conventionally in decision theory $P(B/A)$ appeared in this formula instead of $P(A^{\square}{\rightarrow}B)$:

'Conventional' decision formula:

$$P(L/S)U(S\&L) + P(\neg L/S)U(S\&\neg L) < P(L/\neg S)U(\neg S\&L)$$
$$+ P(\neg L/S)U(\neg S\&\neg L)$$

but it became apparent that this would not do. As the slogan has it, the probability of a counterfactual conditional is not a conditional probability. I can illustrate why with a caricature of a hypothesis mooted by R. A. Fisher. Perhaps smoking does not cause lung cancer; rather the observed probabilistic dependence of lung cancer on smoking arises entirely because both are the result of some gene that is prevalent in the population. Then it might well be the case that $P(L/S) \gg P(S/\neg L)$, but it would not make sense to give up smoking if one loved it in order to avoid lung cancer. To keep the example simple I shall

---

[37] In *Nature's Capacities and their Measurement* the difference between causal-law claims and capacity claims is taken as a difference in levels of modality.

suppose that there is no other cause of lung cancer besides the two possible causes, smoking and the gene.

Since on the 'Fisher' hypothesis the probabilistic dependence between S and L is due entirely to the fact that each is itself dependent on the gene, the dependence between them should disappear if we condition on the presence or absence of the gene. This led causal decision theorists to substitute the partial conditional probabilities $P(L/\pm S\pm G)$ for $P(L/\pm S)$, depending on whether I do indeed have the gene or not ($G = $ I have the smoking/lung cancer gene). If, as we might expect, I have no idea at all whether I have the gene, then I should average over $P(L/\pm S\pm G)$, where the weights for the average would reasonably be based on the frequency with which G appears in the population: $P(+G)$, $P(\neg G)$. In case we can make the additional assumption that the only bearing that the gene has on my utility is through smoking and lung cancer,[38] this line of reasoning results in

Causal decision formula:

$$[P(L/S\&G)P(G) + P(L/S\&\neg G)P(\neg G)]U(S\&L) + [P(\neg L/S\&G)P(G)$$
$$+ P(\neg L/S\&\neg G)P(\neg G)]U(S\&\neg L) < [P(L/\neg S\&G)P(G)$$
$$+ P(L/\neg S\&\neg G)P(\neg G)]U(\neg S\&L) + [P(\neg L/\neg S\&G)P(G)$$
$$+ P(\neg L/\neg S\&\neg G)P(\neg G)]U(\neg S\&\neg L).^{[39]}$$

In the case when G is independent of S ($P(\pm G/\pm S) = P(\pm G)$), this formula reduces to the 'conventional' formula.

Notice that the difference $P([S^{\square}\to L]/\pm G) - P([\neg S^{\square}\to L]/\pm G)$ is given by $P(L/S\&\pm G)P(\pm G) - P(L/\neg S\&\pm G)P(\pm G)$. This latter formula is a direct analogue to Heckman's formula for the causal/counterfactual difference for values. Hold fixed the other causes of the effect in question and see what difference occurs when the targeted cause varies on its own; except that in this case we look not to the difference in values of the effect as the cause varies but rather to the difference in probabilities. I shall by extension call this the probabilistic causal/counterfactual difference. It is clearly not defined whenever S and G are not variation-free; when it is defined and they are variation free, we can also by analogy take the formula to provide a measure of the probabilistic causal contribution of S to L given G or given $\neg G$.[40]

---

[38] So that $U(\pm S\pm L\pm G) = U(\pm S\pm L)$.

[39] When there is more than one common cause involved, the usual generalization of this formula conditions on the state descriptions over the common causes, weighted with the probabilities with which each state description obtains.

[40] In the linear models assumed in section 16.3 in this chapter, the coefficients of each variable are assumed to be functionally independent of the values of all variables, so relativization analogous to the relativization to $+G$ and $\neg G$ here was not necessary. The assumption here analogous to that in section 16.3 would be that S's contribution to L is the same in the presence and in the absence of G.

Like the value-based causal/counterfactual difference this too is more like the counterfactual difference we look for in a Galilean experiment than the implementation-specific difference that might occur in real cases. The particular example chosen tends to obscure this point (as did many others focused on in the early days of causal decision theory). In our case we have only one other cause on the *tapis* and it is unlikely to be changed by any method by which we might come to stop smoking. But suppose that the way in which I will be brought, or bring myself, to stop smoking has some chance of altering whether I have the relevant gene or not. In that case, if we assume that the causal contributions of separate factors are additive, a better formula for the implementation-specific probabilistic counterfactual difference might be[41] (letting $cc(A, B/C)$ stand for the causal contribution of A to B in the presence of C):

$$P([S^\square{\rightarrow}L]/ \pm G) - P([\neg S^\square{\rightarrow}L]/ \pm G)$$
$$= cc(S, L/\neg G) \times P([S^\square{\rightarrow}\neg G]/ \pm G) + [cc(S, L/G)$$
$$+ cc(G, L/S)]P([S^\square{\rightarrow}G]/ \pm G)$$

I offer this formula as an illustration to make a specific point. Behind the story is a small causal model based on the little story I told about smoking, the gene and lung cancer plus the assumption that contributions from separate causes combine additively. And that buys us some advance. But it does not eliminate the counterfactuals altogether. We still need a model involving the implementation variables and the relation to the system to calculate the probability of the remaining counterfactuals. The second model in cases like this will often be far more *ad hoc* and involve far more local knowledge than the one that models the basic system itself.

The overall point of this discussion, however, is that causal decision-theories typically employ a measure that depends entirely on the causal contribution of the action in question. But what is needed, as in policy deliberations in general, is a formula that involves implementation-specific counterfactuals across the range of implementations that might in fact obtain – i.e. 'genuine' counterfactuals.

16.6   Conclusion

I have called many of the counterfactuals of current interest in economics (and in philosophy) 'impostors' because they are generally not answers to the genuine 'What if . . .?' questions of policy and evaluation. Instead they provide a tool

---

[41] I offer this as a plausible example. Whether it is the 'correct' formula or not will, as I have argued, depend on the details of the causal model; and, as I have also already noted, we do not yet have very good prescriptions for getting from the great variety of different kinds of models we employ to methods of evaluating the various different kinds of implementation-neutral and implementation-specific counterfactuals we may need for policy.

for measuring causal contributions in very special kinds of situations – the 'Galilean experiments'. I began with genuine counterfactuals. For purposes of planning and evaluation we need answers to a variety of 'What if . . .?' questions, both implementation-specific questions and implementation-neutral ones. But I have now come full circle. Despite claims of RCT advocates to the contrary, the best way to evaluate these counterfactuals is via a good causal model. And how do we construct an appropriate causal model for answering genuine 'What if . . .?' questions about a given situation? Learning the causal contributions of the relevant factors from Galilean models will be a huge help here, so long as we keep in mind all the strictures about external validity. So, impostor counterfactuals can play a role in answering genuine 'What if . . .?' questions, albeit a very indirect one. But only a role – they cannot provide the real answer.

Bibliography

Adams, P., Hurd, M. D., McFadden, D., Merill, A. and Ribeiro, T. 2003, 'Healthy, Wealthy and Wise? Tests for direct causal paths between health and socioeconomic paths between health and socioeconomic status', *Journal of Econometrics*, 3–56.

Alexandrova, A. 2005, *Connecting Models to the Real World: Game Theory in Action*, unpublished PhD Dissertation, University of California, San Diego.

Anscombe, E. 1971, 'Causality and Determination', reprinted in Sosa, E. and Tooley, M. (eds.) (1993) *Causation*, Oxford: OUP.

Assane, D. and Grammy, A. 2003, 'An Assessment of Growth and Inequality Causality Relationship', *Applied Economics Letters*, 10, 871–3.

Atkinson, A. B. 1987, 'On the Measurement of Poverty', *Econometrica*, 55, 749–64.

1998, *Poverty in Europe*, Oxford: Blackwell.

Balke, A. and Pearl, J. 1995, 'Counterfactuals and Policy Analysis in Structural Models', in P. Besnard and S. Hanks (eds.), *Uncertainty in Artificial Intelligence 11*, San Francisco, CA: Morgan Kaufmann, 11–18.

Bechtel, W. and Abrahamsen, A. forthcoming, 'Phenomena and Mechanisms: Putting the Symbolic, Connectionist, and Dynamical Systems Debate in Broader Perspective', in R. Stainton (ed.), *Contemporary Debates in Cognitive Science*, Oxford: Blackwell.

Berkovitz, J. 2000, 'The Many Principles of the Common Cause', *Reports on Philosophy*, 20, 51–83.

Bridgman, P. W. 1927/1980 *The Logic of Modern Physics*, New York: Arno Press.

Buchdahl, G. 1969, *Metaphysics and the Philosophy of Science*, Oxford: Blackwell.

Card, D. and Krueger, A. 1995, *Myth and Measurement: the New Economics of the Minimum Wage*, Princeton: Princeton University Press.

Cartwright, N. 1979, 'Causal Laws and Effective Strategies', *Nous*, 13, 419–37. Also published in Cartwright (1983).

1983, *How the Laws of Physics Lie*, Oxford: Oxford University Press.

1989, *Nature's Capacities and their Measurement*, Oxford: Clarendon Press.

1995, 'Causal Structures in Econometric Models', in Little, Daniel (ed.), *The Reliability of Economic Models*, Dordrecht: Kluwer Academic Publishers.

1997, 'What is a Causal Structure', in R. McKim Vaughn, and Stephen, P. Turner (eds.), *Causality in Crisis? Statistical Methods and the Search for Causal Knowledge in the Social Sciences*, Indiana: University of Notre Dame Press, 1997.

1998, 'Capacities', in J. B. Davis, D. Wade Hands and U. Mäki (eds.), *The Handbook of Economic Methodology*, Cheltenham: Edward Elgar.

1999, *The Dappled World: a Study of the Boundaries of Science*, Cambridge: Cambridge University Press.

'Causal Diversity and the Markov Condition', *Synthèse*, 121–2, 3–27.

2002, 'Causation, Thick and Thin', University of Nottingham Lecture and unpublished manuscript, London School of Economics.

Cartwright, N. and Jones, M. 1991, 'How to Hunt Quantum Causes', *Erkenntnis* 35, 205–31.

Cartwright, N. forthcoming, 'In Praise of the Representation Theorem', to appear in a special issue of *Dialectica* in honour of Patrick Suppes.

Cat, J. forthcoming, 'Fuzzy Empiricism and Fuzzy-set Causality: What is all the Fuzz about?', *Philosophy of Science*.

Cooley, T. and LeRoy, S. 1985, 'Atheoretical Macroeconomics: a Critique', *Journal of Monetary Economics*, 16, 283–308.

Daly, M. and Wilson, M. 1999, 'An Evolutionary Psychological Perspective on Homicide', in D. Smith and M. Zahn (eds.), *Homicide Studies: a Sourcebook of Social Research* Thousand Oaks, Calif.: Sage Publications, 58–71 (available at: http://psych.mcmaster.ca/dalywilson/chapter5.pdf).

Dowe, P. 2000, *Physical Causation*, Cambridge: Cambridge University Press.

Engle, R., Hendry, D. and Richard, J. F. 1983, 'Exogeneity', *Econometrica*, 51, 277–304.

Fennell, D. 2005a, *A Philosophical Analysis of Causality in Econometrics*, unpublished doctoral dissertation, University of London.

Fennell, D. 2005b, 'Identification in Econometrics and "Possible" Experiments', Causality, Probability and Rationality Conference, Institute of Advanced Study, University of Bologna, Italy, May 6–7 2005.

Finlay, L. and Gough, B. (eds.) 2003, *Reflexivity: a Practical Guide for Researchers in Health and Social Sciences*, Oxford: Blackwell.

Fleming, J. 1998, *Historical Perspectives on Climate Change*, Oxford: Oxford University Press.

Friedman, M. 1953, 'The Methodology of Positive Economics', in *Essays in Positive Economics*, Chicago: Chicago University Press.

Frigg, R. 2000, 'Examples of Non-Contiguous Causation', unpublished manuscript, Centre for the Philosophy of Natural and Social Science, London School of Economics.

Frisch, R. 1938 'Autonomy of Economic Relations', reprinted in D. Hendry and M. Morgan (eds.), 1995, *The Foundations of Econometric Analysis*, Cambridge: Cambridge University Press.

Galavotti, M. C. 2005, 'Plurality in Causality', presented at the Seventh Meeting of the Pittsburgh-Konstanz Colloquium in the Philosophy of Science, May 2005, Konstanz, Germany.

Galison, P. 1997, *Image and Logic: a Material Culture of Microphysics*, Chicago: Chicago University Press.

Galles, D. and Pearl, J. forthcoming, *An Axiomatic Characterization of Causal Counterfactuals*, Technical Report (R-250), prepared for Foundations of Science, Dordrecht: Kluwer.

Geary, D. C. 1996, 'Sexual Selection and Sex Differences in Mathematical Abilities', *Behavioral and Brain Sciences*, 19, 229–84 (available at: http://www.missouri.edu/~psycorie/GearyBBS96.htm).

Glymour, C. 1980, *Theory and Evidence*, Princeton, NJ: Princeton University Press.

Glymour, C., Scheines, R., Spirtes, P. and Kelly, K. 1987, *Discovering Causal Structure*, New York: Academic Press.

Granger, C. 1969, 'Investigating Causal Relations by Econometric Models and Cross-Special Methods', *Econometrica*, 37, 424–38.

   1980, 'Testing for Causality: a Personal Viewpoint', *Journal of Economic Dynamics and Control*, 2, 329–52.

Guala, F. 2005, *The Methodology of Experimental Economics*. Cambridge: Cambridge University Press.

Haavelmo, T. 1944, 'The Probability Approach in Econometrics', *Econometrica*, 12, supplement, iii–vi, 1–115.

Hall, N. and Paul, L. A. 2003, 'Causation and Pre-emption', in P. Clark and K. Hawley (eds.), *Philosophy of Science Today*, Oxford: Clarendon Press, pp. 100–30.

Hamilton, J. 1997, 'Measuring the Liquidity Effect', *American Economic Review*, 87, 80–97.

Harper, W. L., Stalnaker, R. and Pearce, G. (eds.) 1981, *Ifs: Conditionals, Belief, Decision, Chance and Time*, Dordrecht: Reidel.

Hausman, D. 1992, *The Inexact and Separate Science of Economics*, Cambridge: Cambridge University Press.

   1998, *Causal Asymmetries*, Cambridge: Cambridge University Press.

Hausman, D. and Woodward, J. 1999, 'Independence, Invariance, and the Causal Markov Condition', *British Journal for the Philosophy of Science*, 50, 521–83.

   2003, 'Modularity and the Causal Markov Condition', unpublished manuscript.

   2004, 'Modularity and the Causal Markov Condition: a Restatement', *British Journal for the Philosophy of Science*, 55, 147–61.

Heckman, J. 2001, 'Econometrics, Counterfactuals and Causal Models,' Keynote Address International Statistical Institute, Seoul, Korea.

Hempel, C. G. 1966, *Philosophy of Natural Science*, Englewood Cliffs, NJ: Prentice-Hall.

Hendry, D. 2004, *Causality and Exogeneity in Non-stationary Economic Time Series*, Causality: Metaphysics and Methods Technical Report, CTR 18-04, Centre for Philosophy of Natural and Social Science, London School of Economics.

Hendry, D. and Morgan, M. 1995, *The Foundations of Econometric Analysis*, Cambridge: Cambridge University Press.

Hesslow, G. 1976, 'Discussion: Two Notes on the Probabilistic Approach to Causality', *Philosophy of Science*, 43, 290–2.

Hitchcock, C. 2001, 'The Intransitivity of Causation Revealed in Equations and Graphs', *Journal of Philosophy*, 98, 273–99.

Holland, P. W. and Rubin, D. B. 1988, 'Causal Inference in Retrospective Studies', *Evaluation Review*, 12, 203–31.

Hoover, K. 1990, 'The Logic of Causal Inference', *Economics and Philosophy*, 6, 207–34.

   1991, 'The Causal Direction between Money and Prices', *Journal of Monetary Economics*, 27, 381–423.

   2001, *Causality in Macroeconomics*, Cambridge: Cambridge University Press.

Kahneman, D. and Tversky, A. 1979, 'Prospect Theory: an Analysis of Decision Under Risk', *Econometrica*, 47, 263–91.

LeRoy, S. 2003, 'Causality in Economics', unpublished manuscript, University of California, Santa Barbara.

2004, *Causality in Economics*, Causality: Metaphysics and Methods Technical Reports, CTR 20/04, Centre for Philosophy of Natural and Social Science, London School of Economics.

Lessing, G. E. 1759 [1967], *Abhandlungen über die Fabel*, Stuttgart: Philipp Reclam.

Lewis, D. 1973, 'Causation', *Journal of Philosophy*, 70, 556–67.

Lieberson, S. 1992, 'Small N's and Big Conclusions: an Examination of the Reasoning in Comparative Studies Based on a Small Number of Cases', in C. Ragin and H. Becker (eds.), *What is a Case? Exploring the Foundations of Social Inquiry?* Cambridge: Cambridge University Press.

Lucas, R. E. 1981, 'Econometric Policy Evaluation: A Critique', in *Studies in Business Cycle Theory*, Oxford: Basil Blackwell.

Lucas, R. E., 1981, *Studies in Business-Cycle Theory*, Cambridge, MA: MIT Press.

Lundberg, M. and L. Squire, 2003, 'The Simultaneous Evolution of Growth and Inequality', *Economic Journal* 113(487), 326–34.

Macaulay, D. 1988, *The Way Things Work*, Boston: Houghton Mifflin.

Mackie, J. L. 1974, *The Cement of the Universe: a Study of Causation*, Oxford: Clarendon Press.

Mäki, U. 1992, 'On the Method of Isolation in Economics', *Poznán Studies in the Philosophy of the Sciences and the Humanities*, 26, 319–54.

Marmot, M. 2004, *Status Syndrome: How Your Social Standing Directly Affects Your Health and Life Expectancy*, London: Bloomsbury.

Menger, C. 1883 [1963], *Untersuchungen über die Methode der Sozialwissenschaften und der Politischen Oekonomie Insbesondere*, Leipzig: Duncker & Humblot, trans. *Problems of Economics and Sociology*, Urbana: University of Illinois Press.

Menzies, P. and Price, H. 1993, 'Causation as a Secondary Quality', *British Journal for the Philosophy of Science*, 44, 187–203.

Mill, J. S. 1836 [1967], 'On the Definition of Political Economy and on the Method of Philosophical Investigation in that Science', reprinted in *Collected Works of John Stuart Mill*, vol. IV, Toronto: University of Toronto Press.

1843 [1973], 'On the Logic of Moral Sciences', a chapter from *A System of Logic*, reprinted in *Collected Works of John Stuart Mill*, vols. VII–VIII, Toronto: University of Toronto Press.

Mitchell, S. 2003, *Biological Complexity and Integrative Pluralism*, Cambridge: Cambridge University Press.

Morgan, M. 1990, *The History of Econometric Ideas*, Cambridge: Cambridge University Press.

Morgan, M. and Morrison, M. (Eds.) 1999, *Models as Mediators*, Cambridge: Cambridge University Press.

Morrison, M. 1995, 'Capacities, Tendencies and the Problem of Singular Causes', *Philosophy and Phenomenological Research*, 55, 163–8.

Oreskes, N. (forthcoming), 'The Scientific Consensus on Climate Change: How Do We Know We're Not Wrong?' to appear in J. DiMento (ed.), *Climate Change*, Cambridge, MA: MIT Press.

Pearl, J. 1995, 'Causal Diagrams and Empirical Research', *Biometrica*, 82, 669–710.

2000a, *Causality: Models, Reasoning and Inference*, Cambridge: Cambridge University Press.

2000b, 'The Logic of Counterfactuals in Causal Inference (Discussion of "Causal Inference Without Counterfactuals" by A. P. Dawid),' in *Journal of American Statistical Association*, 95, 450, 428–35.

2002, 'Causal Modelling and the Logic of Science', LSE Lakatos lecture.

Pearl, J. and Verma, T. 1991, 'A Theory of Inferred Causation', in J. A. Allen, R. Fikes and E. Sandewall (eds.), *Principles of Knowledge, Representation and Reasoning*, San Mateo: Morgan Kaufmann.

Pissarides, C. 1992, 'Loss of Skill During Unemployment and the Persistence of Unemployment Shocks', *Quarterly Journal of Economics*, 107, 1371–91.

Plott, C. R. 1991, 'Will Economics Become an Experimental Science?', *Southern Economic Journal*, 57, 901–19.

Price, H. 1991, 'Agency and Probabilistic Causality', *British Journal for the Philosophy of Science*, 42, 157–76.

Ragin, C. 1998, 'The Logic of Qualitative Comparative Analysis', *International Review of Social History*, 43, supplement 6, 105–24.

Redhead, M. 1987, *Incompleteness, Nonlocality and Realism: a Prolegomenon to the Philosophy of Quantum Mechanics*, Oxford: Oxford University Press.

Reiss, J. 2002, *Causal Inference in the Abstract or Seven Myths About Thought Experiments*, Causality: Metaphysics and Methods Technical Reports, CTR 03/02, Centre for Philosophy of Natural and Social Science, London School of Economics.

2003, *Practice Ahead of Theory: Instrumental Variables, Natural Experiments and Inductivism in Econometrics*, Causality: Metaphysics and Methods Technical Reports, CTR 12/03, Centre for Philosophy of Natural and Social Science, London School of Economics.

(forthcoming (a)), 'Causal Instrumental Variables and Interventions', *Philosophy of Science*.

(forthcoming (b)), *Taming Error in Economics*, London: Routledge.

Reiss, J., and Cartwright, N. 2003, *Uncertainty in Econometrics: Evaluating Policy Counterfactuals*, Causality: Metaphysics and Methods Technical Report, CTR 11–03, Centre for the Philosophy of Natural and Social Science, London School of Economics.

2004, 'Uncertainty in Econometrics: Evaluating Policy Counterfactuals', in P. Mooslechner, H. Schuberth and M. Schürtz (eds.), *The Role of Truth and Accountability in Policy Advice*, Cheltenham: Edward Elgar.

Russell, B. 1913, 'On the Notion of Cause', *Proceedings of the Aristotelian Society*, 13, 1–26.

Salmon, W. C. 1984, *Scientific Explanation and the Causal Structure of the World*, Princeton: Princeton University Press.

Shafer, G. 1996, *The Art of Causal Conjecture*, Cambridge, MA: MIT Press.

Simon, H. A. 1957a, 'Causal Order and Identifiability', in *Models of Man*, New York: Wiley.

1957b, 'Spurious Causation: a Causal Interpretation' in *Models of Man*, New York: Wiley.

Smith, V. L. 1991, *Papers in Experimental Economics*, Cambridge: Cambridge University Press.

Sober, E. 1988, *Reconstructing the Past*, Cambridge, MA: MIT Press.

1999, 'Instrumentalism Revisited', *Critica*, 91, 3–39.

2001, 'Venetian Sea Levels, British Bread Prices, and the Principle of the Common Cause', *British Journal for the Philosophy of Science*, 52, 331–46.

Sosa, E. and Tooley, M. (eds.) 1993, *Causation*, Oxford: Oxford University Press.

Spirtes, P., Glymour, C. and Scheines, R. 1993, *Causation, Prediction and Search*, New York: Springer-Verlag.

Spirtes, P., Meek, C. and Richardson, T. 1996, *Causal Inference in the Presence of Latent Variables and Selection Bias*, Technical Report CMU-77-Phil., Pittsburgh: Carnegie Mellon University.

Spohn, W. 2001, 'Bayesian Nets Are All There Is to Causal Dependence', in D. Costantini, M. C. Galavotti and P. Suppes (eds.), *Stochastic Causality*, Stanford, CA CSLI Publications.

Suarez, M. 2004, 'An Inferential Conception of Scientific Representation', *Philosophy of Science*, 71, 767–79.

Sugden, R. 2000, 'Credible Worlds: The Status of Theoretical Models in Economics', *Journal of Economic Methodology*, 7, 1–31.

Suppes, P. 1970, *A Probabilistic Theory of Causality*, Amsterdam: North Holland.

Swanson, N. and Granger, C. 1997, 'Impulse Response Functions Based on Causal Approach to Residual Orthogonalization in Vector Autoregressions', *Journal of American Statistical Association*, 92, 357–67.

Swoyer, C. 1991, 'Structural Representation and Surrogative Reasoning', *Synthèse*, 87, 449–508.

Thomson, J. J. 1977, *Acts and Other Events*, London: Cornell University Press.

Williams, B. 1985, *Ethics and the Limits of Philosophy*, Cambridge, MA: Harvard University Press.

Woodward, J. 1997, 'Explanation, Invariance and Intervention', *Philosophy of Science*, 64, S26–S41.

2000a, 'Causation and Manipulation', unpublished manuscript, California Institute of Technology.

2000b, 'Explanation and Invariance in the Special Sciences', *British Journal for the Philosophy of Science*, 51, 197–254.

2003, *Making Things Happen: a Causal Theory of Explanation*, Oxford: Oxford University Press.

Worrall, J. 2002, 'What Evidence in Evidence-Based Medicine?' *Philosophy of Science*, 69, S316–S330.

Index